Keys to the Sciences

Islamic Philosophy, Theology and Science

TEXTS AND STUDIES

Edited by

Hans Daiber
Anna Akasoy
Emilie Savage-Smith

VOLUME 112

The titles published in this series are listed at *brill.com/ipts*

Keys to the Sciences

(Maqālīd al-ʿulūm)
A Gift for the Muzaffarid Shāh Shujāʿ on the Definitions of Technical Terms

Edited with an Introduction by

Gholamreza Dadkhah
Reza Pourjavady

TUTA SUB AEGIDE PALLAS · 1683 ·

BRILL

LEIDEN | BOSTON

Cover illustration: The first page of a copy of *Maqālīd al-ʿulūm*, preserved in National Library of Turkey in Ankara (no. 6790/5, fol. 146[b]).

The Library of Congress Cataloging-in-Publication Data is available online at http://catalog.loc.gov
LC record available at http://lccn.loc.gov/2019058034

Typeface for the Latin, Greek, and Cyrillic scripts: "Brill". See and download: brill.com/brill-typeface.

ISSN 0169-8729
ISBN 978-90-04-42335-0 (hardback)
ISBN 978-90-04-42336-7 (e-book)

This book is printed on acid-free paper and produced in a sustainable manner.

Contents

Acknowledgements

Our deepest gratitude goes to the following people who helped us in the process of edition of this work. Amir Hosein Pourjavady introduced to us the work in the first place. Nasrollah Pourjavady helped us acquire a manuscript of this work from the British Library. İhsan Fazlıoğlu helped us acquire the two other manuscript copies of the work from libraries in Turkey. We would like to thank Sabine Schmidtke with whom we consulted on various editorial matters in the course of preparation of this edition. We also benefited from consultation with Annabel Keeler, Abdollah Anvar, Asadollah Fallahi, Behruz Safarzadeh and Abbas Goudarznia. Arham Moradi and Ahmad-Reza Rahimi-Riseh kindly provided us with some relevant sources. Taro Mimura created the two figures of the work in Chapter Fourteen. Hinrich Biesterfeldt kindly read the introduction and helped us with his insightful comments. We are also grateful for the constructive comments of Hans Daiber and the anonymous reviewer. Farzaneh Sharifi assisted us in the process of transcription. Isabel Miller kindly read the introduction and meticulously edited its English. Finally, we would like to thank Leila Rahimi Bahmany for her comments and assistance in the course of revision of the introduction.

Introduction

1 A Review of the Previous Scholarship

The Swiss orientalist, Charles Rieu (d. 1902), was the first modern scholar to examine the *Maqālīd al-ʿulūm* (Keys to the Sciences).[1] In his *Supplement to the Catalogue of the Arabic Manuscripts in the British Museum*, published in 1894, Rieu included a description of MS British Library Or. 3143, which contained a work titled *Maqālīd al-ʿulūm*. The work he described is on definitions of technical terms in twenty-one sciences.[2] The name of the author is not mentioned anywhere in the text. Nevertheless, the manuscript bears a prefixed title, preportedly by a later hand, attributing the work to Abū l-Faḍl ʿAbd al-Rahmān Jalāl al-Dīn al-Suyūṭī (d. 911/1505). Rieu, however, rejected this attribution on the basis that the work was dedicated to the Muzaffarid Shāh Shujāʿ (r. 759–786/1358–1384), who reigned well before Suyūṭī was born. Instead, Rieu suggested that the work might have been written by Mīr Sayyid Sharīf Jurjānī (d. 816/1413), who is known to have been appointed by Shāh Shujāʿ as a lecturer at the Madrasa of the Dār al-Shifāʾ in Shiraz. According to Rieu, this conjecture is supported by the striking verbal similarities found between certain passages in the *Maqālīd* and Jurjānī's well-established work on the subject, *al-Taʿrīfāt*.[3]

The alleged attribution of the work to Jurjānī was reiterated by Jurji Zaydan (d. 1914) in his *Tārīkh ādāb al-lugha al-ʿarabiyya*, first published in Cairo between 1329/1911 and 1332/1914. Zaydan referred to the manuscript of the *Maqālīd* preserved in the British Library. However, he does not seem to have examined the manuscript personally, and the information he provided was based on the catalogue.[4] The British musicologist and Arabist, Henry George

1 *Maqālīd* is the plural of *miqlīd* or *miqlād*, both of which mean key.

2 See Charles Rieu, *Supplement to the Catalogue of the Arabic Manuscripts in the British Museum* (London: The British Museum, 1894), p. 487, no. 715. The catalogue is available online: https://ia802609.us.archive.org/1/items/supplementtocataoobrituoft/supplementtocataoobrituoft.pdf.

3 As we shall discuss, the similarities between the two works are because of their common sources. See below, pp. 19–21.

4 Jurji Zaydan, *Tārīkh ādāb al-lugha al-ʿarabiyya* (Cairo: Dār al-Hilāl, 1957), vol. 3, p. 252. Based on the information provided by Zaydan, Āqā Buzurg Ṭihrānī included the *Maqālīd al-ʿulūm* in his *al-Dharīʿa ilā taṣānīf al-Shīʿa*. Attributing the work to Jurjānī, Āqā Buzurg refers to MS British Library Or. 3143. See Āqā Buzurg Ṭihrānī, *al-Dharīʿa* (Beirut: Dār al-Aḍwāʾ, 1403–1406/1983–1986), vol. 8, p. 12. Decades later, Khayr al-Dīn al-Ziriklī also attributed this work to Jurjānī in his *al-Aʿlām* (Beirut, Dār al-ʿIlm li-l-Malāyīn, 2002), vol. 5, p. 7.

 | DOI:10.1163/9789004423367_002

Farmer (d. 1965), also attributed the work to Jurjānī, first in his article on the ʿŪd that he wrote for the *Encyclopedia of Islam* in 1938,[5] and then in his *The Sources of Arabian Music*, published in 1940, where he included the *Maqālīd al-ʿulūm* among the Arabic works that dealt with music. Farmer, who seems to have examined the manuscript at the British Library, described the work as a 'pocket encyclopedia of twenty-one sciences'.[6] Concurrently with Farmer, Carl Brockelmann (d. 1956) added the *Maqālīd* to the list of works by Jurjānī and took the attribution made by Rieu for granted, ignoring any uncertainty in this regard.[7] Following Brockelmann, any later scholars who compiled bibliographies of Jurjānī's works included the *Maqālīd* among them.[8]

In 2004, Muḥammad Ibrāhīm ʿUbāda published an edition of the *Maqālīd al-ʿulūm* based on the manuscript of the British Library. ʿUbāda dismissed the attribution of the work to Jalāl al-Dīn al-Suyūṭī on the basis of its content, including its dedication to Shāh Shujāʿ.[9] Remaining completely silent about its attribution to Jurjānī, ʿUbāda speculated that the work might have been composed by a certain Shams al-Dīn Muḥammad b. Ḥasan al-Suyūṭī. ʿUbāda's supposition is based on Jalāl al-Dīn al-Suyūṭī's *Bughyat al-wuʿāh*, where Shams

5 See Henry George Farmer, "ʿŪd', *Encyclopedia of Islam*, vol. 8 (Leiden: Brill, 1938), pp. 985–988(988).

6 See Henry George Farmer, *The Sources of Arabian Music: An Annotated Bibliography of Arabic Manuscripts Which Deal with the Theory, Practice, and History of Arabian Music* (Bearsden: Issued privately by the author, 1940), p. 58, no. 296. Following Farmer, Amnon Shiloah also listed the *Maqālīd* among the Arabic works which deal with music and he attributed it to Jurjānī. See Amnon Shiloah, *The Theory of Music in Arabic Writings (C. 900–1900): Descriptive Catalogue of Manuscripts in Libraries of Europe and U.S.A.* (Munich: RISMB, 1979), pp. 99–100.

7 Carl Brockelmann, *Geschichte der arabischen Litteratur*, 2 vols. (Leiden: Brill, 1943–1949), vol. 2, p. 280, no. 3.

8 See Mahdī Shakībāniyā and Reza Pourjavady, 'Kitābshināsī-i Sayyid Sharīf-i Jurjānī', *Maʿārif* 19 3 (2003): 134–192 (177); Boris A. Rosenfeld and Ekmeleddin İhsanoğlu, *Mathematicians, Astronomers & Other Scholars of Islamic Civilization and Their Works (7th–19th c.)* (Istanbul: Research Center for Islamic History, Art and Culture, 2003), pp. 265–266; Josef van Ess, 'Jorjāni, Zayn-al-Din Abu'l-Hasan ʿAli', *Encyclopædia Iranica*, vol. 15/1, 2009, pp. 21–29 (28); idem, *Die Träume der Schulweisheit: Leben und Werk des ʿAlī b. Muḥammad al-Ǧurǧānī (gest. 816/1413)* (Wiesbaden: Harrassowitz Verlag, 2013), p. 86. Some scholars, however, have remained uncertain about this attribution. See Sonja Brentjes, 'Courtly Patronage of the Ancient Sciences in Post-Classical Islamic Societies', *al-Qanṭara* 29 2 (2008): 403–436 (426); Amir Hosein Pourjavady, 'Jurjānī, ʿAlī b. Muḥammad: Mūsīqī', *Dāʾirat al-maʿārif-i buzurg-i Islāmī*, ed. Kāẓim Mūsawī Bujnūrdī, vol. 17 (Tehran: Markaz-i Dāʾirat al-Maʿārif-i Buzurg-i Islāmī, 1388 Sh./2009), pp. 714–715.

9 Jalāl al-Dīn al-Suyūṭī (attrib.), *Maqālīd al-ʿulūm fī l-ḥudūd wa-l-rusūm*, ed. Muḥammad Ibrāhīm ʿUbāda (Cairo: Maktabat al-Ādāb, 1424/2004), pp. 6–11.

al-Dīn al-Suyūṭī is introduced as an Arabist acquainted with a number of sciences (*ʿārifan bi-ʿiddat funūn*), who died in 808/1405–1406.[10] This is, however, unlikely; on the front page of the manuscript in the British Library, the author is introduced as Abū l-Faḍl Jalāl al-Dīn al-Suyūṭī and not Shams al-Dīn al-Suyūṭī. Moreover, in the short biographical entry on Shams al-Dīn al-Suyūṭī, nothing is reported about his settling in Shiraz or the neighbouring regions. Hence there is no reason to believe that he was connected to the court of Shāh Shujāʿ.

2 Patronage at the Court of Shāh Shujāʿ

The only significant piece of information about the author of the *Maqālīd* is his dedication of the work to Shāh Shujāʿ. Therefore, in order to know more about the author, an examination of what can be known of the court of Shāh Shujāʿ, and the scholars associated with it, is a requirement.

In 759/1358, Shāh Shujāʿ together with his brother Shāh Maḥmūd (r. 759–776/1358–1375) deposed their father, Amīr Mubāriz al-Dīn (r. 736–759/1335–1358) and divided his territory between themselves. Whereas Shāh Shujāʿ ruled Shiraz, Shāh Maḥmūd reigned in Isfahan. The *Maqālīd* must have been written in Shiraz sometime after 759/1358 and before Shāh Shujāʿ's death in 786/1384 with the exception of a period between 765/1363 and 767/1366, during which Shāh Maḥmūd took control of almost all the lands of Shāh Shujāʿ.

Being himself a poet with a *dīwān*, albeit a minor one, Shāh Shujāʿ favoured the poets at his court.[11] Ḥāfiẓ (d. 792/1390) was the most famous poet associated with Shāh Shujāʿ's court. Other poets who eulogised him and enjoyed his patronage included Rukn al-Dīn Ṣāʾin Hirawī (d. 764/1362–1363),[12] ʿUbayd

10 Ibid., pp. 11–12. Cf. Jalāl al-Dīn al-Suyūṭī, *Bughyat al-wuʿāh fī ṭabaqāt al-lughawiyyin wa-l-nuḥāh* (Cairo: Maṭbaʿat ʿĪsā l-Bābī al-Ḥalabī, 1964), vol. 1, p. 91.

11 For Shāh Shujāʿ's poems and the preface (*dībāja*) to his *dīwān* by Saʿd Unsī, see Qāsim Ghanī, *Tārīkh-i ʿaṣr-i Ḥāfiẓ: Baḥth dar āthār u afkār u aḥwāl-i Ḥāfiẓ* (Tehran: Zawwār 1383 Sh./2004), pp. 344–355; Muḥammad Qazwīnī, *Yāddāshthā-yi Qazwīnī*, ed. Īraj Afshār (Tehran: Intishārāt-i Dānishgāh-i Tihrān, 1388 Sh./2009), vol. 9, pp. 1–14. On Shāh Shujāʿ as a mediocre poet, see ibid., vol. 9, p. 4, fn. 2. For more information about Shāh Shujāʿ's poems, see Jawād Basharī, 'Barrasī-i awḍāʿ-i adabī-i Fārs dar sada-yi hashtum u manābiʿ-i īn barrasī', *Āyina-yi Pazhūhish* 17 1 (1385 Sh./2006): 69–80 (97).

12 For the poems of Rukn al-Dīn Ṣāʾin Hirawī in praise of Shāh Shujāʿ, see Rukn al-Dīn Ṣāʾin Hirawī, *Dīwān-i Ṣāʾin-i Hirawī* (Patna: Muʾassasa-yi Taḥqīqāt u Tatabbuʿāt dar Zabān u Adab-i ʿArabī u Fārsī, 1959), pp. 160–199.

Zākānī (d. 772/1371),[13] ʿImād al-Dīn Faqīh Kirmānī (d. 773/1371),[14] Iftikhār al-Dīn Dāmghānī (d. 775/1373–1374),[15] Salmān Sāwajī (d. 778/1376),[16] Jalāl al-Dīn Ṭabīb Shīrāzī (d. 785/1383),[17] Jahān-Malik Khātūn (d. after 784/1382),[18] ʿAḍud al-Dīn Ṣarrāf Yazdī (d. after 790/1388),[19] and ʿIzz al-Dīn Muṭahhar Ḥusaynī (fl. 782/1380).[20]

Little is known about Shāh Shujāʿ's education. He seems to have studied literary sciences with Muʿīn al-Dīn Yazdī (d. 789/1387–1388), the author of the chronicle *Mawāhib-i ilāhī*.[21] On these same subjects, he also benefited occa-

---

13 For the poems of ʿUbayd Zākānī in praise of Shāh Shujāʿ, see ʿUbayd Zākānī, *Kulliyyāt-i ʿUbayd-i Zākānī*, ed. ʿAbbās Iqbāl and Parwīz Atābakī (Tehran: Iqbāl, 1340 Sh./1961), vol. 1, pp. 4–5, 7, 21–22, 38–40.

14 On ʿImād al-Dīn Faqīh Kirmānī, see J.T.P. de Bruin, "ʿImād al-Dīn ʿAlī, Faḳīh-i Kirmānī', in *Encyclopaedia of Islam. 2nd Ed.*, suppl., pp. 414–416. For his poems in praise of Shāh Shujāʿ, see ʿImād al-Dīn Faqīh Kirmānī, *Dīwān-i ʿImād-i Faqīh-i Kirmānī*, ed. Rukn al-Dīn Humāyūn (Tehran: Ibn Sīnā, 1348 Sh./1969), pp. 10–11, 32–33, 316–317, 339–342.

15 Iftikhār al-Dīn Dāmghānī dedicated one of his *qaṣīdas*, titled *Rawshanāyī-nāma* to Shāh Shujāʿ. See Iftikhār al-Dīn Dāmghānī and Ḥasan Dhulfaqārī (ed.), 'Rawshanāyī-nāma', in *Nāmwāra-yi duktur Maḥmūd Afshār*, ed. Īraj Afshār (Tehran: Bunyād-i Mawqūfāt-i Duktur Maḥmūd Afshār, 1373 Sh./1994), vol. 8, pp. 4300–4305; cf. Reza Pourjavady, 'The Legacy of Aḍud al-Dīn al-Ījī: His Works and His Students', in Ayman Shihadeh and Jan Thiele, ed., *Philosophical Theology in Islam: Later Ashʿarism East and West* (Leiden and Boston: Brill, 2020), pp. 337–370 (360).

16 For the poems of Salmān Sāwajī in praise of Shāh Shujāʿ, see Salmān Sāwajī, *Dīwān-i Salmān-i Sāwajī*, ed. Manṣūr Mushfiq with an introduction by Taqī Tafaḍḍulī (Tehran: Ṣafī ʿAlī-Shāh, 1367 Sh./1988–1989), pp. 610–613.

17 For the poem(s) of Jalāl al-Dīn Ṭabīb Shīrāzī in praise of Shāh Shujāʿ, see his *Dīwān*, ed. Nasrollah Pourjavady (Tehran: Farhangistān-i Zabān u Adab-i Fārsī, 1389 Sh./2010), pp. 20–22, 108.

18 For the poems of Jahān-Malik Khātūn in praise of Shāh Shujāʿ, see Jahān-Malik Khātūn, *Dīwān-i kāmil-i Jahān-Malik Khātūn, qarn-i hashtum-i hijrī*, ed. Pūrāndukht Kāshānī Rād and K. Aḥmadnizhād (Tehran: Zawwār, 1374 Sh./1995), pp. 1–2, 7–8. On Jahān-Malik Khātūn, see Dominic Parviz Brookshaw, 'Jahān-Malek Ḵātūn', *Encyclopædia Iranica*, vol. 14, pp. 383–385; Domenico Ingenito, 'Jahan Malik Khatun: Gender, Canon, and Persona in the Poems of a Premodern Persian Princes', in Alireza Korangi, Hanadi Al-Samman, Michael C. Beard, ed., *The Beloved in Middle Eastern Literature: The Culture of Love and Languishing* (London and New York: I.B. Tauris, 2018), pp. 177–212, 330–337.

19 For ʿAḍud al-Dīn Ṣarrāf Yazdī's poems in praise of the shah, see ʿAḍud al-Dīn Ṣarrāf Yazdī, *Dīwān-i ʿAḍud*, ed. ʿAlī Riḍā Qūjazāda (Tehran: Kitābkhāna, Mūzi u Markaz-i Asnād-i Majlis-i Shūrā-yi Islāmī, 1389 Sh./2010), pp. 6–8, 14–16, 28–29, 31–32, 87, 162, 177.

20 For the poem of ʿIzz al-Dīn Muṭahhar Ḥusaynī in praise of Shāh Shujāʿ, see Ghanī, *Tārīkh-i ʿaṣr-i Ḥāfiẓ*, pp. 257–259.

21 On Shāh Shujāʿ's study with Muʿīn al-Dīn Yazdī, see Saʿīd Nafīsī's introduction to his edition of Muʿīn al-Dīn Yazdī's *Mawāhib-i ilāhī dar tārīkh-i Āl-i Muẓaffar*, vol. 1 (Tehran: Kitābkhāna u Chāpkhāna-yi Iqbāl, 1326 Sh./1947), pp. 6–9.

sionally from the teaching of Aḍud al-Dīn Ījī (d. 756/1356)[22] and Qiwām al-Dīn Shīrāzī (d. 772/1370).[23] From the texts which Shāh Shujāʿ reportedly studied with his teachers, such as Ibn Ḥājib's *Mukhtaṣar* and from Arabic poems attributed to him, one may infer that he knew Arabic. He even wrote a short piece in Arabic titled *On the Nobility of Knowledge and the Merit of Learning and Teaching* (*fī sharaf al-ʿilm wa-faḍl al-taʿallum wa-l-taʿlīm*).[24] It was reported that literary debates were held at his court.[25] The author of the *Maqālīd* also indicates that a group of scholars was affiliated to Shāh Shujāʿ's court, and the shah himself took an active part in debates on various subjects with the scholars. Thus, the shah was certainly able to benefit from learned works in Arabic dedicated to him, including the *Maqālīd al-ʿulūm* (for the list of other works in prose dedicated to Shāh Shujāʿ, see Appendix 1).

One of the most prominent scholars present at the court of Shāh Shujāʿ was Iftikhār al-Dīn Dāmghānī who studied in Tabriz with ʿAḍud al-Dīn Ījī and apparently moved to Shiraz after the collapse of the Ilkhanate in 736/1336. His numerous works indicate his mastery of learning in a variety of fields such as Qurʾānic exegesis, legal theory, rational theology, rhetoric, dialectic, logic, and philosophy.[26] Dāmghānī was also a poet with the pen name Iftikhār. At Shāh Shujāʿ's request, Dāmghānī wrote a treatise, titled *Ḥall mā lā yanḥall* (Resolving the Unresolvable), addressing a controversial question in rhetoric related to one of the Qurʾānic verses of the challenge (Q. 2:23, *faʾtū bi-sūratin min mithlih*).[27] Another distinguished scholar of the time was Bahāʾ al-Dīn ʿAbd al-

22 In 754/1353, when Ījī went to the camp of Amīr Mubāriz al-Dīn, the latter asked Ījī to teach his son. Thus, he taught Shāh Shujāʿ some parts of Ibn Ḥājib's commentary on Zamakhsharī's *Mufaṣṣal*. See Muʿīn al-Dīn-i Yazdī, *Mawāhib-i ilāhī*, pp. 241–245.

23 In his *Tārīkh-i Āl-i Muẓaffar*, Maḥmūd Kutubī indicated that Shāh Shujāʿ started attending the lectures of Qiwām al-Dīn Shīrāzī after he recaptured Shiraz in 767/1366. He was reading ʿAḍud al-Dīn Ījī's commentary on Ibn Ḥājib's (d. 646/1249) 'abridgement' of his *Muntahā-l-suʾl wa-l-amal fī ʿilmay al-uṣūl wa-l-jadal* with Shīrāzī. See Maḥmūd Kutubī, *Tārīkh-i Āl-i Muẓaffar*, ed. ʿAbd al-Ḥusayn Nawāʾī (Tehran: Intishārāt-i Amīr Kabīr, 1364 Sh./1985), p. 95. On Qawām al-Dīn Shīrāzī, see Muʿīn al-Dīn Junayd al-Shīrāzī, *Shadd al-izār fī ḥaṭṭ al-awzār ʿan zuwwār al-mazār*, ed. Muḥammad Qazwīnī and ʿAbbās Iqbāl (Tehran: Chāpkhāna-yi Majlis, 1328 Sh./1949), pp. 84–87.

24 For the edition of this essay, see Ghanī, *Tārīkh-i ʿaṣr-i Ḥāfiẓ*, pp. 346–349.

25 See below in Appendix 1, the description of *Ḥall mā lā yanḥall* by Iftikhār al-Dīn Dāmghānī.

26 Among Dāmghānī's works, one should mention his commentary on Nāṣir al-Dīn Bayḍāwī's (d. 685/1286) *Ṭawāliʿ al-anwār*, the work from which the author of the *Maqālīd* quotes long passages.

27 For more information on Iftikhār al-Dīn Dāmghānī and his works, see Pourjavady, 'The Legacy of Aḍud al-Dīn al-Ījī', pp. 358–360.

Ṣamad Baḥrābādī Isfarāyinī (d. 786/1384–1385). A descendant of the Kubrawī Sufi shaykh, Saʿd al-Dīn Ḥamūya (d. 650/1260), Baḥrābādī composed works on logic, rational theology, and jurisprudence.[28] He was a close companion of Shāh Shujāʿ in Shiraz. He was given the honour of sitting in the front rank of the debating sessions (*majālis*) at the court.[29]

The two scholars, mentioned above, were primarily experts in the religious and literary sciences. However, they were potentially capable of writing the *Maqālīd*. Evidently, there were more scholars of this kind connected to the court. Besides scholars in various fields of religion and literature, Shāh Shujāʿ also seems to have patronized physicians. His enthusiasm for promoting medical studies and medical care in his kingdom is evident from the construction of a hospital in Shiraz called the Dār al-Shifāʾ, which seems to have been established during his reign and on his orders.[30] Furthermore, an anonymous physician wrote *al-Tawḍīḥ fī sharḥ al-Talwīḥ* and dedicated it to the shah. Another physician of the time was ʿAlī b. Ḥusayn Anṣārī, known as Zayn al-ʿAṭṭār (d. 776/1374–1375), who composed an extensive medical work titled *Ikhtiyārāt-i badīʿī* and dedicated it to Amīr Mubāriz al-Dīn's wife in 770/1368–1369.[31] He was one of the court physicians. Another was Jalāl al-Dīn Aḥmad Ṭabīb Shīrāzī, who was also a distinguished poet.[32] Yet another physician in Shiraz, Manṣūr b. Muḥammad Shīrāzī, composed a Persian medical treatise, titled *Kifāyat al-Mujāhidiyya* in 783/1381–1382 and dedicated it to Shāh Shujāʿ's son, the prince Mujāhid al-Dīn Zayn al-ʿĀbidīn (r. 786–789/1384–1387).[33] The number of physicians in Shiraz at this time was certainly not restricted to these individuals.[34] Bearing in mind that the chapter on medicine

28 Among the works he reportedly composed were *Qawānīn al-manṭiq*, a commentary on the *ʿAqāʾid* of ʿAḍud al-Dīn al-Ījī and *Makārim al-sharīʿa*. See Ibid., pp. 459–460.

29 According to Shīrāzī, he was the sulṭān's companion and among his duties was to sit at the front of his *majālis* (*al-taṣaddur fī al-majālis*). See Shīrāzī, *Shadd al-izār*, p. 459.

30 It has been assumed that the hospital was located on the same site where the Buyid ʿAḍud al-Dawla, who ruled Shiraz between 338/949 and 366/977, constructed a hospital. See Ḥasan Khūbnaẓar and Jaʿfar Muʾayyid Shīrāzī, *Tārīkh-i Shīrāz: az āghāz tā ibtidā-yi salṭanat-i Karīm Khān-i Zand* (Tehran: Sukhan, 1380 Sh./2001–2002), p. 597.

31 This work has been partially edited. See ʿAlī b. Ḥusayn Anṣārī Shīrāzī, *Ikhtiyārāt-i badīʿī*, ed. Muḥammad Taqī Mīr (Tehran: Shirkat-i Dārūyī-i Pakhsh-i Rāzī, 1371 Sh./1992–1993).

32 On Jalāl Ṭabīb Shīrāzī, see Nasrollah Pourjavady's introduction to his *Dīwān*, pp. 1–47.

33 Numerous manuscript copies of this work are extant—for instance, MS Marʿashī 11298.

34 For instance, Mubārakshāh Bukhārī, who dedicated his commentary on Ṣafī al-Dīn Urmawī's *al-Adwār* to Shāh Shujāʿ (see Appendix), was primarily interested in medicine. See Mubārakshāh Bukhārī, *Tarjuma-yi sharḥ-i Mubārakshāh-i Bukhārī bar Adwār-i Urmawī dar mūsīqī*, ed. and Persian trans. Sayyid ʿAbdullāh Anwār (Tehran: Farhangistān-i Hunar, 1392 Sh./2013), p. 18.

is the most extensive chapter of the *Maqālīd*, it could have been one of these physicians who composed the work.

Although Shāh Shujāʿ spent most of his time in Shiraz, it is also possible that the *Maqālīd* was dedicated to him in another place such as Tabriz, which he conquered in 776/1375. He stayed in Tabriz for only a few months. However, it was long enough to attract scholars of the city to the learned shah.[35] There is also the possibility that a scholar from another city composed the work. For instance, Sayyid Ḥaydar Āmulī (d. after 787/1385) dedicated his voluminous exegetical work, *al-Muḥīṭ al-aʿẓam*, to Shāh Shujāʿ in 777/1376.[36] Apparently, Āmulī was living in Najaf at the time, and he dedicated the work to Shāh Shujāʿ during a short visit to Shiraz.[37] Thus, it could be that like Āmulī, the author of the *Maqālīd* was a scholar living in another city, who dedicated the work to the shah while he was staying in Shiraz.

In 779/1378, on his way back to Shiraz from Yazd, Shāh Shujāʿ met Sayyid Sharīf Jurjānī and being impressed by his knowledge, invited him to lecture at the Madrasa of the Dār al-Shifāʾ. Jurjānī was active in Shiraz for the next ten years (779–789/1378–1387), the first seven years of which were the last ones of Shāh Shujāʿ's reign.[38] Although we cannot rule out the possibility that Jurjānī was the author of the *Maqālīd*, there are reasons to doubt this. First, none of the works whose attribution to him is certain was dedicated to Shāh Shujāʿ, and so in this respect, the *Maqālīd* would be an exception.[39] Second, Jurjānī's works were all widely popular among scholars. It is likely that Jurjānī himself, as well as his students, promoted his works. In contrast, *Maqālīd al-ʿulūm* barely, if ever, received any scholarly attention. Third, as we will discuss below, the structure and content of Jurjānī's *Taʿrīfāt* differ from those of the *Maqālīd* such that it is unlikely that they had a common author.

---

35 Faḍl Allāh Astarābādī narrates a story of a certain ʿIzz al-Dīn from Tabriz who was willing to dedicate one of his works to the shah in this period. See Shahrzad Bashir, *Fazlallah Astarabadi and the Hurufis* (Oxford: One World, 2005), p. 25.

36 See below, Appendix 1, p. 30.

37 On Sayyid Ḥaydar Āmulī's biography, see Josef van Ess, 'Ḥaydar-i Āmulī', *Encyclopaedia of Islam, 2nd Ed.*, Supplement volume XII, pp. 363–365.

38 On Jurjānī's first meeting with Shāh Shujāʿ and his residence in Shiraz in this period, see van Ess, *Die Träume der Schulweisheit*, pp. 25–34.

39 The only known authority to whom Jurjānī dedicated his works is the Timurid prince, Iskandar Mīrzā (r. 811–817/1415). Jurjānī's magnum opus, a commentary on Ījī's *Mawāqif* is dedicated to him. What is more, Jurjānī wrote a Persian treatise in response to the questions set him by Iskandar Mīrzā. See the edition of this work by Maḥmūd Yazdī Muṭlaq (Fāḍil), in *Muḥaqqiq-Nāma: Maqālāt-i taqdīm-shuda bi ustād duktur Mahdī Muḥaqqiq*, ed. Jūyā Jahānbakhsh and Bahāʾ al-Dīn Khurramshāhī (Tehran: Anjuman-i Āthār u Mafākhir-i Farhangī, 1380 Sh./2001–2002), vol. 2, pp. 1389–1447.

3 The Structure and the Sources of the *Maqālīd*

The *Maqālīd al-ʿulūm* contains definitions of 1862 technical terms in twenty-one sciences and consists of a preface, an introduction, and twenty-one chapters (*abwāb*, sg. *bāb*).

In the preface, the author says that he collected the material for this book in his early youth either from the lips of his masters in their teachings and disputation sessions or from the popular books on these sciences. He further says that he was encouraged by one of his friends (*baʿḍ ikhwānī fī l-dīn wa-aḥad khullānī min ahl al-yaqīn*) to compile the scattered notes and make them into a single work. The author provides some information about the way he structured the chapters: in each chapter, he first defines the subject matter of the science in question and then defines its technical terms. Finally, he explains the various objectives of the book: firstly, to present the purpose of each field of learning and familiarize the student with its respective technical terms, which may be of assistance to a beginner in studying it. Secondly, to demonstrate that certain technical terms have different meanings in different disciplines.[40] Thirdly, to help the reader to more mastery in disputations and debates.

In the introduction, the author first distinguishes between the principles (*mabādiʾ*) and the premises (*muqaddamāt*) of each science. Then he explains the different ways of defining things, using technical philosophical terms. He differentiates between the complete definition (*al-ḥadd al-tāmm*), the incomplete definition (*al-ḥadd al-nāqiṣ*), the complete description (*al-rasm al-tāmm*) and the incomplete description (*al-rasm al-nāqiṣ*). The author further defines the essence (*al-māhiyya*) and explains the correspondence between a proper conceptual knowledge of a thing and its external essence. In this discussion, he draws heavily upon Nāṣir al-Dīn Bayḍāwī's (d. 685/1286) *Ṭawāliʿ al-anwār*, referring to its author on one occasion as 'the *qāḍī*' without specifying his name or the title of his work.[41]

The twenty-one chapters devoted to technical terms of the sciences are as follows: (1) exegesis (*al-tafsīr*), (2) tradition (*al-ḥadīth*), (3) jurisprudence (*al-fiqh*), (4) legal theory (*uṣūl al-fiqh*), (5) rational theology (*uṣūl al-kalām*), (6) dialectic (*al-jadal*), (7) syntax (*al-naḥw*), (8) morphology (*al-ṣarf*), (9) rhetoric (*al-maʿānī wa-l-bayān*), (10) prosody (*al-ʿarūḍ*), (11) logic (*al-manṭiq*), (12) philosophy (*al-ḥikma*), (13) astronomy (*al-hayʾa*), (14) geometry (*al-handasa*),

---

40 In fact, throughout the work, some terms appear several times in different chapters with different definitions. See below, Table 2.

41 See Nāṣir al-Dīn al-Bayḍāvī, *Ṭawāliʿ al-anwār min maṭāliʿ al-anḍār*, ed. ʿAbbās Sulaymān (Beirut: Dār al-Jīl, 1411/1991), pp. 56–59.

(15) arithmetic (*al-ḥisāb*), (16) finance (*al-istīfāʾ*), (17) music (*al-mūsīqā*), (18) astrology (*al-nujūm*), (19) medicine (*al-ṭibb*), (20) ethics (*al-akhlāq*) and (21) Sufism (*al-taṣawwuf*). The sequence of the sciences presented in this work does not necessarily represent the author's system of classification; he claims that he has mixed the rational and religious sciences (*ʿulūm al-ʿaql wa-sharʿ*) in a manner which is in harmony with nature (*yuwāfiqu al-ṭabʿ*). The first ten chapters of the book deal with the religious and literary sciences (exegesis, tradition, jurisprudence, legal theology, rational theology, dialectic, syntax, morphology, rhetoric and prosody), followed by chapters on the five rational sciences (logic, philosophy, astronomy, geometry and arithmetic). The sequence of the final six sciences does not follow any particular classification. However, it does not seem to be altogether arbitrary either, in particular, the order of the last three chapters of the book, which deal with the state of the human being as an individual: first, medicine, which deals with the states of body and its health, is considered; then ethics, which deals with the states that characterize a person; and finally Sufism, which deals with the states that occur internally and externally to a human being as the consequence of the correction of their actions (*al-ʿilm bi-l-aḥwāl al-mawrūtha min taṣḥīḥ al-aʿmāl ẓāhiran wa-bāṭinan*).

In each chapter, the first definition is the definition of the subject matter. For some terms, several definitions are presented, of which the first seems to be the one preferred by the author. Furthermore, the chapters are mostly divided into sections, which are marked by the author under the simple heading of *faṣl*. One can determine the specific subjects of most of these sections. For instance, the chapter on jurisprudence contains the following typical subdivisions: (1) rules for religious rites (*fiqh al-ʿibādāt*), (2) family law (*fiqh al-munākaḥāt*), (3) rules covering social relations (*fiqh al-muʿāmalāt*), and (4) criminal law (*fiqh al-jināyāt*). However, there are some sections which do not seem to have an overarching theme. For instance, the chapter on music contains three sections: (1) on composition (*fī l-taʾlīf*), (2) on frets, strings, and parts of the ʿūd and related matters (*fī l-dasātīn wa-l-awtār wa-ajzāʾ al-ʿūd wa-mā yatbaʿuhā*), (3) rhythm (*fī l-īqāʿāt*). However, at the end of the third section on rhythm, some terms are included which are ostensibly irrelevant to its subject (ex. *al-ṭanīnī, al-lawnī, al-taʾlīfī*).

As mentioned above, the author of the *Maqālīd* says that he made use of both oral and written sources for the definitions of terms. Nevertheless, it seems that in some chapters his reliance on written sources was more substantial than in others. In what follows, the chapters of the *Maqālīd* are discussed with respect to their possible written sources (for a more detailed study, see Appendix 2).

On Qur'ānic exegesis, the author might have consulted Bayḍāwī's *Minhāj al-wuṣūl fī sharḥ al-uṣūl* and Abū Muḥammad al-Ḥusayn Baghawī's *Maʿālim al-tanzīl* (*Tafsīr al-Baghawī*).[42] However, we were unable to detect most of the definitions in earlier works. Efforts to unearth the written sources for the chapter on jurisprudence were inconclusive. On legal theory, the author again might have consulted Bayḍāwī's *Minhāj al-wuṣūl fī sharḥ al-uṣūl*.[43] The chapter on the tradition (*ḥadīth*) is sophisticated in terms of the material presented. Many of the definitions were probably derived from Sharaf al-Dīn Ṭayyibī's (d. 743/1342) *Khulāṣat al-uṣūl fī maʿrifat ḥadīth al-Rasūl*.[44] On rational theology, a number of the definitions were taken from Shams al-Dīn Samarqandī's (d. 722/1322) *al-Ṣaḥāʾif al-ilāhiyya*.[45] On dialectic, the author extracted certain definitions from Samarqandī's *Risāla fī ādāb al-baḥth*.[46] Nevertheless, the

42 As far as we could determine, the definitions given for *al-ʿām* and *al-mubīn* in the chapter on Qur'ānic exegesis can be found in *Minhāj al-wuṣūl*. See Nāṣir al-Dīn al-Bayḍāwī, *Minhāj al-wuṣūl ilā ʿilm al-uṣūl*, ed. Shaʿbān Muḥammad Ismāʿīl (Beirut: Dār al-Ḥazm, 1429/2008), pp. 121, 139. Similarly, the definitions of *al-tafsīr* and *al-taʾwīl* can be found in Baghawī's *Maʿālim al-tanzīl*. See Abū Muḥammad al-Baghawī, *Tafsīr al-Baghawī*, ed. Muḥammad ʿAbdullāh al-Namir, ʿUthmān Jumʿa al-Ḍaymariyya, and Sulaymān Muslim al-Ḥarash (Riyadh: Dār Ṭayyiba, 1409/1988–1989), vol. 1, p. 46. Moreover, the distinction between *al-tafsīr* and *al-taʾwīl* can be harked back to Shams al-Dīn Āmulī's (d. 753/1352) *Nafāʾis al-funūn*. See Shams al-Dīn Āmulī, *Nafāʾis al-funūn fī ʿarāʾis al-ʿuyūn*, ed. Abū l-Ḥasan Shaʿrānī, 3 vols. (Tehran: Intishārāt-i Islāmiyya, 1377–1379/1957–59), vol. 1, pp. 369–370. For more details, see below, Appendix 2, pp. 32–33.

43 The definitions given for *uṣūl al-fiqh*, *al-ḥukm*, *al-wājib*, *al-mandūb*, and *tarāduf* in the chapter on legal theory can be found in *Minhāj al-wuṣūl*. See al-Bayḍāwī, *Minhāj al-wuṣūl*, pp. 51–56, 82, 85. For more details, see below, Appendix 2, p. 35.

44 See Sharaf al-Dīn al-Ṭayyibī, *al-Khulāṣa fī l-ḥadīth*, ed. Abū ʿĀṣim al-Shawāmī al-Atharī (Cairo: al-Maktaba al-Islāmiyya, 1430/2009). Some of the definitions can be traced back in Ibn Ṣalāḥ Shahrazūrī's (d. 643/1245) *ʿUlūm al-ḥadīth* and Badr al-Dīn Ibn Jamāʿa's (d. 733/1333) *al-Manhal al-rawī*. The latter work was extensively appropriated by Ṭayyibī in his composition of the *Khulāṣa*. See Abū ʿAmr ʿUthmān ibn Ṣalāḥ al-Shahrazūrī, *ʿUlūm al-ḥadīth*, ed. Nūr al-Dīn ʿAtar (Damascus: Dār al-Fikr, 1406/1986); Badr al-Dīn Ibn Jamāʿa, *al-Manhal al-rawī fī mukhtaṣar ʿulūm al-ḥadīth al-nabawī*, ed. Muḥyī al-Dīn ʿAbd al-Raḥmān Ramaḍān (Damascus: Dār al-Fikr, 1406/1985–1986).

45 More than twenty definitions of the technical terms in the chapter on rational theology are derived from *maqṣad* 1, *qism* 1 and 3 and *maqṣad* 2 of *al-Ṣaḥāʾif al-ilāhiyya*. See Shams al-Dīn al-Samarqandī, *al-Ṣaḥāʾif al-ilāhiyya*, ed. Aḥmad ʿAbd al-Raḥmān al-Sharīf (Kuwait City: al-Fallāḥ, 1985), pp. 73–164, 253–495.

46 The author of the *Maqālīd* seems to have used twelve definitions, which he mentions at the beginning of the treatise. For Samarqandī's *Risāla fī ādāb al-baḥth*, see MS Majlis-i Shūrā-yi Islāmī 9006/1, pp. 3–8. Cf. Mehmet Kadri Karabela, 'The Development of Dialectic and Argumentation Theory in Post-Classical Islamic Intellectual History', PhD Dissertation, McGill University, Montreal, 2010, pp. 266–270. A critical edition of this treatise, together with its translation, is currently under preparation by Walter Edward Young.

chapter contains many technical terms which could not be harked in any earlier sources.

On syntax, the author used Ibn Ḥājib's (d. 646/1249) manual of syntax, *al-Kāfiya*.[47] On this subject, he also seems to have consulted Tāj al-Dīn Isfarāyinī's (d. 684/1285–1286) *al-Lubāb fī ʿilm al-iʿrāb*.[48] On morphology, the author has extracted the definitions from Ibn Ḥājib's *al-Shāfiya*.[49] He may have used this textbook with the commentary on it by Fakhr al-Dīn Jārbardī (d. 746/1345–1346).[50] As regards rhetoric (*al-maʿānī wa-l-bayān*), many of the definitions given can be found in Jalāl al-Dīn Khaṭībī Qazwīnī's (d. 739/1338) *al-Īḍāḥ* and Sirāj al-Dīn Sakkākī's (d. 626/1229) *Miftāḥ al-ʿulūm*, Sharaf al-Dīn Ṭayyibī's *al-Tibyān fī al-bayān* and Ibn Nāẓim al-Dimashqī's (d. 686/1287–1288) *al-Miṣbāḥ*.[51] In the chapter on prosody, he appropriated most of the terms and definitions that he supplied from Yaḥyā b. ʿAlī Khaṭīb Tabrīzī's (d. 502/1109) *al-Kāfī fī ʿilmay al-ʿarūḍ wa-l-qawāfī*, and Ibn Sarrāj al-Shantarīnī's (d. 508/1114) *al-Kāfī fī ʿilm al-qawāfī*.[52]

On logic, he extensively benefited from Najm al-Dīn Kātibī's (d. 675/1277) *al-Shamsiyya* together with Quṭb al-Dīn Rāzī's (d. 766/1365) commentary on it.[53] For the definitions of the terms related to philosophy, again the author

---

47 See Ibn al-Ḥājib, *al-Kāfiya fī ʿilm al-naḥw wa-l-shāfiya fī ʿilmay al-taṣrīf wa-l-khaṭṭ*, ed. Ṣāliḥ ʿAbd al-ʿAẓīm al-Shāʿir (Cairo: Maktabat al-Ādāb, 1431/2010).

48 One occasion on which he seems to have consulted *al-Lubāb* was for the definition of '*kalama*', where an extra definition, based on this source, is provided: *qīla: hiya al-lafẓ al-mawḍūʿ li-l-maʿnā mufradan*. See Tāj al-Dīn al-Isfarāʾinī, *al-Lubāb fī ʿilm al-iʿrāb*, ed. Shawqī al-Maʿarrī (Beirut: Maktabat Lubnān Nāshirūn, 1996), p. 13.

49 See Ibn al-Ḥājib, *al-Kāfiya fī ʿilm al-naḥw wa-l-shāfiya fī ʿilmay al-taṣrīf wa-l-khaṭṭ*, pp. 59–106. For more details, see below, Appendix 2, pp. 41–42.

50 At least one of the definitions in this chapter corresponds to a definition in Jārbardī's commentary on the *Shāfiya*: that for *al-āla*. See Fakhr al-Dīn al-Jārbardī, 'Sharḥ al-Shāfiya', in *Majmūʿa al-Shāfiya fī ʿilmay al-taṣrīf wa-l-khaṭṭ*, ed. Muḥammad ʿAbd al-Salām Shāhīn (Beirut: Dār al-Kutub al-ʿIlmiyya, 1971), vol. 1, p. 320. For more details, see below, Appendix 2, p. 41.

51 See Jalāl al-Dīn al-Khaṭīb al-Qazwīnī, *al-Īḍāḥ fī ʿulūm al-balāgha: al-maʿānī wa-l-bayān wa-l-badīʿ* (Beirut: Dār al-Kutub al-ʿIlmiyya, 1424/2003); Sirāj al-Dīn al-Sakkākī, *Miftāḥ al-ʿulūm*, ed. Nuʿaym Zarzūr (Beirut: Dār al-Kutub al-ʿIlmiyya, 1987); Sharaf al-Dīn al-Ṭayyibī, *al-Tibyān fī l-bayān*, ed. Yaḥyā Murād (Beirut: Dār al-Kutub al-ʿIlmiyya, 1425/2004); Ibn Nāẓim al-Dimashqī, *al-Miṣbāḥ fī l-maʿānī wa-l-bayān wa-l-badīʿ*, ed. Ḥusnī ʿAbd al-Jalīl Yūsuf (Cairo: Maktabat al-Ādāb, 1409/1989). For more details, see below, Appendix 2, pp. 42–45.

52 See Yaḥyā b. ʿAlī al-Khaṭīb al-Tabrīzī, *al-Kāfī fī l-ʿarūḍ wa-l-qawāfī*, ed. al-Ḥassānī Ḥasan ʿAbdullāh (Cairo: Maktabat al-Khānajī, 1994/1415), pp. 143–203. Ibn Sarrāj al-Shantarīnī, *Kitāb al-kāfī fī ʿilm al-qawāfī*, ed. D. ʿAlāʾ Muḥammad Raʾfat (Cairo: Dār al-Ṭalāʾiʿ, 2003), pp. 35–37. For more details, see below, Appendix 2, p. 45–50.

53 See Quṭb al-Dīn al-Rāzī, *Taḥrīr al-qawāʿid al-manṭiqiyya fī sharḥ al-Risāla al-Shamsiyya*,

made use of Samarqandī's *al-Ṣaḥā'if al-ilāhiyya*, and this time his adoption of the material was even more substantial than in the chapter on rational theology. Moreover, the author derived a number of definitions from Ibn Sīnā's (d. 428/1037) *Risālat al-Ḥudūd*.[54] *Risālat Ḥudūd al-ashyā' wa-rusūmihā* attributed to al-Kindī (d. 256/873), Ibn Sīnā's *'Uyūn al-ḥikma*, and Abū 'Abdullāh Khʷārazmī's (fl. 366/976) *Mafātīḥ al-'ulūm* are the other possible sources of this chapter.[55] In chapters thirteen to seventeen, namely the chapters on astronomy, geometry, arithmetic, finance and music, the author of the *Maqālīd* relied heavily on Khʷārazmī's *Mafātīḥ al-'ulūm*.[56] In the chapter on astrology, again he depended considerably on the *Mafātīḥ al-'ulūm*.[57] Nevertheless, he must have consulted other sources, which have eluded us. The impact of the *Mafātīḥ al-'ulūm* can be seen in the chapter on medicine too. However, the main sources for this chapter were Ibn Nafīs' (d. 687/1288) *al-Ṣaydaliyya al-mujarraba* and Ibn Sīnā's *al-Qānūn*.[58] In ethics, all of the terms and the definitions are based on Rāghib Iṣfahānī's (d. 502/1108) *al-Dharī'a ilā makārim al-sharī'a*.[59] Finally, in the chapter on Sufism, he mainly used Abū l-Qāsim

---

ed. Muḥsin Bīdārfar (Qom: Intishārāt-i Bīdār, 1386 Sh./2007), pp. 30, 33, 52, 60, 236, 237, 241, 263, 452, 454–460. For more details, see below, Appendix 2, p. 57–62.

54 At least fifteen definitions in this chapter are identical with those of Ibn Sīnā in his *Risālat al-Ḥudūd*. See Ibn Sīnā, 'Risāla al-Ḥudūd', in 'Abd al-Amīr al-A'ṣam, ed., *Rasā'il manṭiqiyya li-l-falāsifa al-'Arab* (Beirut: Dār al-Manāhil. 1428/2008), pp. 111–153 (138–150). For more details, see below, Appendix 2, p. 57–62.

55 The definition of *al-ḥikma* can be traced back in *'Uyūn al-ḥikma*, the one on *al-falsafa* can be found in *Risālat Ḥudūd* attributed to al-Kindī, and the definition of *al-kumūn* can be harked back in the *Mafātīḥ al-'ulūm*. See Ibn Sīnā. *'Uyūn al-ḥikma*, ed. 'Abd al-Raḥmān Badawī, 2nd Ed. (Beirut: Dār al-Qalam, 1980), p. 16; al-Kindī, 'Risāla al-Ḥudūd', in 'Abd al-Amīr al-A'ṣam, ed., *Rasā'il manṭiqiyya li-l-falāsifa al-'Arab* (Beirut: Dār al-Manāhil), 1428/2008, pp. 59–79; Abū 'Abdullāh Muḥammad b. Aḥmad Khʷārazmī, *Liber Mafâtîh al-olûm: explicans vocabula technica scientiarum tam Arabum quam peregrinorum*, ed. Gerlof van Vloten (Leiden: Brill, 1968), p. 140. For more details, see below, Appendix 2, p. 57–58.

56 The author of the *Maqālīd* drew upon *maqāla* 1, *bāb* 4 and *maqāla* 2, *bāb* 4–7 of the *Mafātīḥ al-'ulūm* for these chapters. See Khʷārazmī, *Liber Mafâtîh al-olûm*, pp. 53–79, 184–209, 215–225, 235–246. For more details, see below, Appendix 2, p. 63–72.

57 Ibid., pp. 225–235.

58 For instance, the definitions of *al-qāshir* and *al-mumallis* were taken from Ibn al-Nafīs' *al-Ṣaydaliyya al-mujarraba*. See Ibn al-Nafīs, *al-Ṣaydaliyya al-mujarraba*. See Ibn al-Nafīs, *al-Ṣaydaliyya al-mujarraba*, ed. Muḥsin 'Aqīl (Beirut: Dār al-Maḥajja al-Bayḍā', 1423/2002), pp. 96–97. The definition of *al-riyāḍa* is taken from Ibn Sīnā's *Qānūn*. See Ibn Sīnā, *al-Qānūn fī l-ṭibb*, ed. Muḥammad Amīn al-Ḍanāwī (Beirut: Dār al-Kutub al-'Ilmiyya, 1420/1999), vol. 1, p. 221.

59 See Abū l-Qāsim al-Rāghib al-Iṣfahānī, *al-Dharī'a ilā makārim al-sharī'a*, ed. Abū Yazīd Abū Zayd al-'Ajamī (Cairo: Dār al-Islām, 1428/2007), pp. 73–74, 96–97, 109, 118, 137–138, 140–141, 160–161, 201, 207, 207, 210–213, 244–245, 261. For more details, see below, Appendix 2, pp. 78–81.

Qushayrī's (d. 465/1072) *al-Risāla* and Abū Bakr Muḥammad Kalābādhī's (d. ca. 380/990) *al-Taʿarruf* (Table 1).[60]

TABLE 1 The sources of the *Maqālīd*

	The subject of the sciences	Sources
1.	Exegesis	Possibly Bayḍāwī's *Minhāj al-wuṣūl* Possibly Baghawī's *Maʿālim al-tanzīl*
2.	Tradition	Ṭayyibī's *Khulāṣa fī maʿrifat al-ḥadīth*
3.	Jurisprudence	Unknown
4.	Legal theory	Possibly Bayḍāwī's *Minhāj al-wuṣūl*
5.	Rational Theology	Samarqandī's *al-Ṣaḥāʾif al-ilāhiyya*
6.	Dialectic	Samarqandī's *Risāla fī ādāb al-baḥth*
7.	Syntax	Ibn Ḥājib's *Kāfiya*
8.	Morphology	Ibn Ḥājib's *Shāfiya* Possibly Jārbardī's commentary on the *Shāfiya*
9.	Rhetoric	Khaṭīb Qazwīnī's *al-Īḍāḥ fī ʿulūm al-balāgha* Sakkākī's *Miftāḥ al-ʿulūm* Possibly Ṭayyibī's *al-Tibyān fī al-bayān* Possibly Ibn Nāẓim's *al-Miṣbāḥ*
10.	Prosody	Khaṭīb Tabrīzī's *al-Kāfī fī ʿilmay al-ʿarūḍ wa-l-qawāfī* Ibn Sarrāj's *al-Kāfī fī ʿilm al-qawāfī* Khʷārazmī's *Mafātīḥ al-ʿulūm*
11.	Logic	Kātibī's *al-Shamsiyya* together with Quṭb al-Dīn al-Rāzī's *Taḥrīr al-qawāʿid al-manṭiqiyya*
12.	Philosophy	Samarqandī's *al-Ṣaḥāʾif al-ilāhiyya* Ibn Sīnā's *Risālat al-Ḥudūd* Possibly al-Kindī's *Risālat al-Ḥudūd* Possibly Ibn Sīnā's *ʿUyūn al-ḥikma* Possibly Khʷārazmī's *Mafātīḥ al-ʿulūm*

60 For example, the definition of *al-waqt* was adopted from Qushayrī's *Risāla*. See Abū l-Qāsim al-Qushayrī, *al-Risāla al-Qushayriyya*, ed. ʿAbd al-Ḥamīd Maḥmūd and Maḥmūd b. Sharīf (Cairo: Dār al-Shaʿb, 1409/1989), p. 130. The definition of *al-uns* can be found in Kalābādhī's *al-Taʿarruf*. See Abū Bakr Muḥammad al-Kalābādhī, *al-Taʿarruf li-madhhab ahl al-taṣawwuf*, ed. Aḥmad Shams al-Dīn (Beirut: Dār al-ʿIlm, 1422/2001–2002), pp. 76–77. For more details, see below, Appendix 2, pp. 81–82.

TABLE 1 The sources of the *Maqālīd* (*cont.*)

	The subject of the sciences	Sources
13.	Astronomy	Khʷārazmī's *Mafātīḥ al-ʿulūm*
14.	Geometry	Khʷārazmī's *Mafātīḥ al-ʿulūm*
15.	Arithmetic	Khʷārazmī's *Mafātīḥ al-ʿulūm*
16.	Finance	Khʷārazmī's *Mafātīḥ al-ʿulūm*
17.	Music	Khʷārazmī's *Mafātīḥ al-ʿulūm*
18.	Astrology	Khʷārazmī's *Mafātīḥ al-ʿulūm*
19.	Medicine	Ibn Nafīs' *al-Ṣaydaliyya al-mujarraba* Ibn Sīnā's *Qānūn* Khʷārazmī's *Mafātīḥ al-ʿulūm*
20.	Ethics	Rāghib Iṣfahānī's *al-Dharīʿa ilā makārim al-sharīʿa*
21.	Sufism	Qushayrī's *Risāla* Kalābādhī's *Taʿarruf*

It should be noted that the author of the *Maqālīd* describes his written sources as 'the works commonly used by scholars' (*naqaltu ʿan kutubin tadāwalūhā*). Our findings confirm that the sources for the *Maqālīd* were to a large extent textbooks and manuals on various subjects. However, there are at least two works that were used to which the above description does not fit, namely Khʷārazmī's *Mafātīḥ al-ʿulūm* and Rāghib Iṣfahānī's *al-Dharīʿa*. None of these two works enjoyed much publicity at the time. About *al-Dharīʿa*, we know at least that it was an object of some attention from the Muzaffarid scholars because it was translated into Persian at the request of Shāh Shujāʿ.[61] However, Khʷārazmī's *Mafātīḥ al-ʿulūm*, to our knowledge, was an obscure source for the scholars of the period.[62]

61 See below the description of *Kunūz al-wadīʿa min rumūz al-Dharīʿa* in Appendix 1, p. 29.

62 According to Bosworth, the work 'was little known in the Islamic world as a whole'. See C.E. Bosworth, 'Some New Manuscripts of al-Khwārazmī's *Mafātīh al-ʿulūm*', *Journal of Semitic Studies* 9 2 (1964): 341–345 (345).

4 *Mafātīḥ al-ʿulūm* as the Model for *Maqālīd al-ʿulūm*

Writing books on definitions of technical terms has a long tradition in the Islamicate world. One can trace this type of writing back to the 2nd/8th century when Jābir b. Ḥayyān (d. ca. 200/815) composed his *Risāla fī l-ḥudūd*.[63] However, the majority of books on technical terms are devoted to a single science or a group of related sciences. *Mafātīḥ al-ʿulūm*, to our knowledge, is a pioneering work in Arabic which deals with the technical terms of various sciences. It is not only the first work of its kind, but also, at least to our knowledge, the only work, other than *Maqālīd al-ʿulūm*, which consists of the technical terms of various sciences, yet deals with each science individually. As mentioned above, the *Mafātīḥ al-ʿulūm* was one of the main sources for the *Maqālīd*. The appropriation of the *Mafātīḥ al-ʿulūm* by the author of the *Maqālīd al-ʿulūm* is quite substantial; numerous definitions in the nine chapters of the *Maqālīd* were taken from the *Mafātīḥ*. However, the *Mafātīḥ al-ʿulūm* was more than a mere source of terms and definitions for the author of the *Maqālīd*. In a close comparison between the two books, we can notice several resemblances between them. The two authors had a similar intention in the composition of their works. Just like the author of the *Maqālīd*, Khʷārazmī in his introduction indicates that certain technical terms have more than one meaning in different disciplines and the *Mafātīḥ* is aimed at distinguishing between the various meanings of the terms within each branch of science. This intention appears to be crucial for both authors, and this is the reason they organised their books based on a wide variety of sciences.

The *Maqālīd* also bears a resemblance to Khʷārazmī's *Mafātīḥ* in terms of the classification of the sciences. Khʷārazmī divides his book into two parts (*maqālāt*, sing. *maqāla*): the first on the Islamic religious sciences (*ʿulūm al-sharīʿa*) and the Arab sciences connected with them, and the second on foreign sciences (*ʿulūm al-ʿajam*). Each part is divided into several sections (*abwāb, sing. bāb*). The subjects of the sections of the two parts are: I. (1) jurisprudence (*al-fiqh*), (2) rational theology (*al-kalām*), (3) syntax (*al-naḥw*), (4) the secretary's art (*al-kitāba*), (5) poetry and prosody (*al-shiʿr wa-l-ʿarūḍ*), (6) history (*al-akhbār*). II. (1) philosophy (*al-falsafa*), (2) logic (*al-manṭiq*), (3) medicine (*al-ṭibb*), (4) arithmetic (*al-arithmāṭīqī*), (5) geometry (*al-handasa*), (6) astro-

63 For the early Arabic works of this genre, see M.A.S. Abdel Haleem, 'Early Islamic Theological and Juristic Terminology: *Kitāb al-ḥudūd fī l-uṣūl*, by Ibn Fūrak', *Bulletin of the School of Oriental and African Studies* 54 1 (1991): 5–41 (5–11).

nomy (*al-hay'a*), (7) astrology (*al-nujūm*), (8) music (*al-mūsīqā*), (9) mechanical devices (*al-ḥiyal*), and (10) alchemy (*al-kīmiyā'*).[64]

Although the author of the *Maqālīd* did not follow Khʷārazmī in dividing his book into two parts, the sequence of the chapters presented in it appears to follow the pattern designed by Khʷārazmī in the *Mafātīḥ*. The first ten chapters deal with religious and Arabic literary sciences and the second ten chapters are those which were called by Khʷārazmī 'foreign sciences'.[65] Nevertheless, the author of the *Maqālīd* introduced some alterations into the structure of his work. In the *Mafātīḥ* the secretary's art (*al-kitāba*) is classified among the sciences of the Arabs next to the literary sciences. Instead, the author of the *Maqālīd* included finance (*al-istīfā'*), which is a science related to the tasks of the secretaries, next to mathematical sciences. Moreover, the author of the *Maqālīd* excluded history from the first part and removed mechanical devices and alchemy from the second. Instead, he added eight other sciences, namely exegesis, tradition, dialectic, morphology, rhetoric, ethics, and Sufism. He also separated legal theory from jurisprudence and, likewise, astrology from astronomy. The final chapter, on Sufism, was meant to be distinguished from the

64 As mentioned above, nine chapters of Khʷārazmī's *Mafātīḥ al-ʿulūm* were more or less appropriated by the author of the *Maqālīd*, namely the chapters on secretary's art, and poetry and prosody from the first part, and the chapters on philosophy, medicine, arithmetic, geometry, astronomy, astrology, and music from the second part. It is notable here that some of these chapters have been translated into European languages. Hence, for any partial or complete translations of the corresponding chapters of the *Maqālīd*, the translations of these parts will bear consultation. E. Wiedemann translated part of the chapter on secretary's art, the chapter on geometry and the chapter on Arithmetic into German. See E. Wiedemann, 'Beiträgen zur Geschichte der Naturwissenschaft x. Zur Technik bei den Arabern', *Sitzungsberichte der physikalisch-medizinischen Sozietät in Erlangen* xxxviii (1906): 307–357; idem, 'Beiträgen zur Geschichte der Naturwissenschaft xxii. Stucke aus den Mafâtîḥ al-ʿUlûm', *Sitzungsberichte der physikalisch-medizinischen Sozietät in Erlangen* xlii (1910): 303–322; 'Beiträgen zur Geschichte der Naturwissenschaft xiv. Über die Geometrie und Arithmetik nach den Mafâtiḥ al-ʿUlûm', *Sitzungsberichte der physikalisch-medizinischen Sozietät in Erlangen* xl (1908): 1–29. Together with W. Müller, Wiedemann also made a German translation of the chapter on music. See E. Wiedemann and W. Müller, 'Beiträgen zur Geschichte der Naturwissenschaft lxvi. Zur Geschichte der Musik', *Sitzungsberichte der physikalisch-medizinischen Sozietät in Erlangen* liv–lv (1922–1923): 7–16. E. Seidel translated the chapter on medicine into German. See E. Seidel, 'Die Medizin in Kitâb Mafâtîḥ al-ʿUlûm', *Sitzungsberichte der physikalisch-medizinischen Sozietät in Erlangen* xlvii (1915): 1–79. Moreover, Farmer translated the chapter on music into English. See Henry George Farmer, 'The Science of Music in the *Mafātīḥ al-ʿulūm*', *Transactions Glasgow University Oriental Society*, xvii (1957–1958): 1–9.

65 For more about the structure of Khʷārazmī's *Mafātīḥ al-ʿulūm*, see C.E. Bosworth, 'A Pioneer Arabic Encyclopedia of the Sciences: al-Khwārazmī's Keys of the Sciences', *isis* 14 (1964): 97–111.

previous groups of sciences. Besides the reasons which have been indicated so far, even the names of the two works are also alike. *Maqālīd*, plural of *miqlād* (or *miqlīd*), means 'keys' just like *mafātīḥ*, plural of *miftāḥ*.

As compared to the *Mafātīḥ*, one can see some improvements in the *Maqālīd*. First of all, the number of sciences discussed are increased in the *Maqālīd*. Second, the author of the *Maqālīd* committed himself to defining the subject matter of each science at the beginning of the chapter devoted to its technical terms. But Khʷārazmī does not commit himself to introduce each science in this manner. Moreover, whereas Khʷārazmī did not hesitate to include names, such as names of the instruments, astronomical tools as well as proper names as entries, the author of the *Maqālīd*, as a rule, excluded all such names as entries.[66] There are also considerable improvements in the *Maqālīd* in terms of definitions. Undoubtedly, some sciences had advanced in the interval between the time of the writing of the *Mafātīḥ* in the tenth century and that of the writing of the *Maqālīd* in the fourteenth century. For instance, *al-Kāfiya* by Ibn Ḥājib surpassed all the previous textbooks of syntax. The same applies to *al-Shamsiyya* by Kātibī in logic. Being aware of some of these advancements, the author of the *Maqālīd* substituted the terms and definitions of the *Mafātīḥ* with the more up-to-date terms and definitions based on the recent handbooks. In other words, the author of the *Maqālīd*, revised, updated and upgraded the *Mafātīḥ al-ʿulūm* to the extent his knowledge allowed him to.[67]

It is also notable that Jurjānī's *Taʿrīfāt* was composed in the same period that the *Maqālīd* was written. However, unlike the *Maqālīd* and the *Mafātīḥ*, *Taʿrīfāt* presents all terms of different sciences together in alphabetical order. Later works on the definitions of technical terms, such as ʿAbd al-Raʾūf al-Manāwī's (d. 1031/1621–1622) *al-Tawqīf ʿalā muhimmāt al-taʿārīf*, Muḥammad ʿAlī Tahānawī's (fl. 1158/1745) *Kashshāf iṣṭilāḥāt al-funūn*, and ʿAbd al-Nabī Nakīrī Hindī's (fl. 1173/1759) *Jāmiʿ al-ʿulūm fī iṣṭilāḥāt al-funūn* (known as *Dustūr al-ʿulamāʾ*), were likewise arranged alphabetically. The *Maqālīd* is the only known work on definitions of technical terms which structurally follows Khʷārazmī's *Mafātīḥ al-ʿulūm*.

---

66 For instance, the astronomical tools and musical instruments, which were introduced in the *Mafātīḥ al-ʿulūm* (pp. 232–235, 236–240), were not included in the *Maqālīd al-ʿulūm*.

67 This was not, however, the case with respect to terms and definitions of a few chapters of the *Maqālīd* including the chapter on mathematical sciences and the chapter on finance, which remained heavily based on the Khʷārazmī's *Mafātīḥ al-ʿulūm*, despite the considerable development of these sciences between the 10th and the 14th centuries.

5 Some Serendipitous Knowledge of the Author

The text of the *Maqālīd* provides us with some information about its author. Apart from the fact that the work is dedicated to Shāh Shujāʿ, its praise of the shah as someone whose questions are sweeter than honey and whose comments and responses are better than pearls and pure gold suggests that the author had at least once witnessed a debating session at the court in which Shāh Shujāʿ was actively involved.[68] However, the author did not regard himself as one of the regular scholarly associates of the court whom he refers to as 'the scholars' (*al-ʿulamāʾ*) and 'the learned men' (*al-fuḍalāʾ*).[69] In fact, the dedication of this work to Shāh Shujāʿ was meant to promote the work among those learned scholars affiliated to the court.[70]

The author was probably at an advanced age when he was completing the work; since he states that the materials used in it were collected back in his early youth.[71] From his praise of Shāh Shujāʿ as the 'revival of the glorious deeds of the Rightly Guided caliphs', one can conclude that he was a Sunnī scholar.[72] His use of Bayḍāwī's *Minhāj al-wuṣūl* in legal theory might be an indication that he was Shāfiʿī in legal matters.

The author's heavy reliance on Khʷārazmī's *Mafātīḥ al-ʿulūm* in the chapters concerning astronomy, geometry, arithmetic, finance, and music suggests that he was not expert in these subjects. Furthermore, the author does not seem to have been interested in the occult sciences, apart from perhaps astrology. Noteworthy is also a lack of use of the works of ʿAḍud al-Dīn Ījī, whose dogmatic and literary writings were authoritative for many scholars of the period. This might be an indication that the author was not a student of Ījī. His extensive appropriation of Shams al-Dīn Samarqandī's *al-Ṣaḥāʾif al-ilāhiyya* in the fields of rational theology and philosophy makes him less likely to be an Ashʿarī theologian.[73] Finally, the author of the *Maqālīd* does not show any interest in Ibn

68 See below, p. 91, line 14–16.
69 See below, p. 91, line 16–17.
70 See below p. 91, line 6–8.
71 See below, p. 90, line 15.
72 See below, p. 91, line 10.
73 On Samarqandī's theological tendency, see Heidrun Eichner, *Towards the Construction of Islamic Orthodoxy: Philosophy in the Post-Avicennian Period and Islamic Theology as Literary Traditions* (Habilitationsschrift, Martin-Luther-Universität Halle-Wittenberg, 2009), pp. 310–314, 379–424; Gholamreza Dadkhah's Persian introduction to Shams al-Dīn Samarqandī's *Science of the Cosmos and the Soul: ʿIlm al-ʾāfāq wa al-ʾanfus* (California, Costa Mesa: Mazda Publishers, 2014), pp. 54–60.

ʿArabī's school of mysticism. It might be that he deliberately avoided consulting ʿAbd al-Razzāq Kāshānī's (d. ca. 736/1335) *Iṣṭilāḥāt al-Ṣūfiyya*, or indeed any other work by the members of this school. Instead, he chose two classical Sufi manuals as his main sources on the subject.

The author of the *Maqālīd* implies that he studied a wide range of disciplines in his youth in an environment in which major scholars were debating with each other. Although he does not specify the city in which he studied, his general description of the environment fits Tabriz better than any other cities of the region. In other words, he might have studied in Tabriz in the last years of the Ilkhanid dynasty, when he could have attended the courses of distinguished scholars in various disciplines such as Fakhr al-Dīn Jārbardī, Sharaf al-Dīn Ṭayyibī and Quṭb al-Dīn Rāzī. These are some significant points of speculation about the author, whose identity remains a mystery unless one day a newly discovered manuscript of the work will reveal it.

6 *Maqālīd al-ʿulūm* and the *Taʿrīfāt*: Similarities and Differences

Contrary to what Rieu stated, relatively speaking, there are few verbal coincidences between the *Maqālīd* and the *Taʿrīfāt*. In terms of structure, unlike the *Maqālīd*, Jurjānī's *Taʿrīfāt* was arranged alphabetically and not based according to the sciences to which the terms belong. The author of the *Maqālīd* presents each term with its different definitions under the various disciplines. In fact, he is most emphatic about distinguishing between the different meanings of terms. In the introduction, he presents some examples of this: for instance, *al-sanad* has two different meanings, depending on whether it is being used in the science of tradition (*ḥadīth*) or that of dialectic. In the course of the book, he presents the two definitions of *al-sanad* under each of the two sciences. In the *Taʿrīfāt*, however, only one of the definitions of *al-sanad* is provided, and that is according to its application in dialectic.[74] In some other instances in the *Taʿrīfāt*, Jurjānī distinguished between different meanings of a term in different sciences, but he was less concerned with the different meanings of terms than the author of the *Maqālīd* was (for some other examples, see Table 2).

---

74 See Sayyid Sharīf al-Jurjānī, *Kitāb al-Taʿrīfāt*, ed. Ibrāhīm al-Abyārī (Cairo: Dār al-Rayyān li-l-Turāth, 1403/1982–1983), p. 161.

TABLE 2 Definitions of some terms with multiple meanings in the *Maqālīd al-ʿulūm* and the *Taʿrīfāt*

Terms	The sciences in which the term is used according to the *Maqālīd*	The sciences in which the term is applied according to the *Taʿrīfāt*
al-basīṭ	1. Geometry; 2. Philosophy	1. [Philosophy]
al-buʿd	1. Music; 2. Sufism	1. [Philosophy]
al-faṣl	1. Prosody; 2. Logic	1. [Logic]; 2. Rhetoric
al-ikhlāṣ	1. Ethics; 2. Sufism	1. [Ethics]; 2. Sufism
al-iʿtirāḍ	1. Rhetoric; 2. Astrology	1. [Rhetoric]
al-ḥāl	1. Rational Theology; 2. Syntax 3. Rhetoric; 4. Sufism	1. Syntax; 2. Sufism
al-ḥumq	1. Medicine; 2. Ethics	–
al-jamāʿa	1. Jurisprudence; 2. Finance; 3. Music	–
al-khāṣṣa	1. Astronomy; 2. Logic	1. [Logic]
al-muttaṣil	1. Tradition; 2. Syntax; 3. Philosophy	–
al-murakkab	1. Medicine; 2. Logic; 3. Philosophy	1. [Logic]
al-rajʿa	1. Finance; 2. Astronomy	1. [Jurisprudence]
al-raʾy	1. Legal Theory; 2. Ethics	
al-takhayyul	1. Prosody; 2. Ethics	–
al-zamān	1. Philosophy; 2. Astronomy	1. [Philosophy]

There is no reason to think that the *Maqālīd* was used in the course of authorship of the *Taʿrīfāt* or *vice versa*. If Jurjānī was the author of the *Maqālīd*, given that it was presumably written before the *Taʿrīfāt*, then we needed an explanation for the fact that it is not used, or even referred to, in the *Taʿrīfāt*. It is also worth noting that, as he says in the introduction, Jurjānī's *Taʿrīfāt* is exclusively based on the books and writings of previous scholars, while the author of the *Maqālīd* explicitly declared that he had made use of certain oral sources as well. For these reasons, it seems unlikely that Jurjānī was the author of the *Maqālīd*. The similarity of the definitions in both works is because they had sources in common.[75] For instance, both these authors benefited from

75 See Shakībāniyā and Pourjavady, 'Kitābshināsī-i Sayyid Sharīf-i Jurjānī', 177; Sonja Brentjes,

Kātibī's *al-Shamsiyya*, Samarqandī's *Risāla fī ādāb al-baḥth*, and Khaṭīb Tabrīzī's *al-Kāfī fī ʿilmay al-ʿarūḍ wa-l-qawāfī*.

7 The Question of the Reception of the Work

Unlike Jurjānī's *Taʿrīfāt*, no evidence suggests that the *Maqālīd* was ever transcribed during the Timurid and Safavid domains and eras. As will be explained later, the extant manuscripts of this work were all produced in the Ottoman empire from the 17th century onwards. The books of the Muzaffarid royal library, where the original manuscript/holograph of the *Maqālīd* was most likely held, were probably dispersed after the collapse of the kingdom. Presumably, some of them eventually entered the libraries of Ottoman territory.[76] However, it seems that none of the major libraries of Istanbul held copies of the *Maqālīd* at least until the early decades of the seventeenth century. Otherwise, Ḥājjī Khalīfa (or Kâtip Çelebi, d. 1068/1657) would surely have mentioned the work in his *Kashf al-ẓunūn ʿan asāmī al-kutub wa-l-funūn*. Nevertheless, based on the location of the extant copies of the work, it is more likely that Ottoman rather than Iranian scholars used the *Maqālīd*.

8 The Present Edition

As mentioned earlier, hitherto only MS British Library Or. 3143 of the *Maqālīd* has been known to modern scholars. However, two further manuscripts have recently surfaced in Turkish libraries.[77] The present edition is based, therefore, on all three manuscripts:

آ = Milli Kütüphane-Ankara 6790/5, fols. 146^b–205^b, 205.130–135.65 mm. 17 lines per page. The hand of the manuscript suggests an 11th/17th-century dating. The copy is executed in *nastaʿlīq* and is fully pointed. Most of the chapter

'Courtly Patronage of the Ancient Sciences in Post-Classical Islamic Societies', 426; van Ess, *Die Träume der Schulweisheit*, p. 86.

76 For instance, MS Haci Selim Aga 902 (held at the Süleymaniye Library) contains the holograph of a Persian translation of Jār Allāh Zamakhsharī's *Rabīʿ al-abrār* by Muḥammad b. Aḥmad Ṣūfī Samarqandī, titled *Nasīm al-Rabīʿ*. Completed in 781/1379, the work is dedicated to Muẓaffar al-Dīn Salghur Shāh, the vizier of Shāh Shujāʿ. From the ornamentation of this manuscript, it seems very likely that it was produced for the royal library.

77 We are grateful to İhsan Fazlıoğlu, who helped us acquire the images of these manuscripts.

and section headings are in red ink, and the entries are usually marked by a red line above them. It lacks catchwords. The manuscript contains marginal notes by a hand similar to the hand of the main text. The notes are quotations from various sources, such as Muḥammad b. Yūsuf Qūnawī Qarapīrī's (fl. 10th/16th century) *Tarjumān al-Ṣiḥāḥ* (fols. 146ᵇ–149ᵃ), an Arabic-Turkish lexicon titled *al-Akhtarī* by Muṣṭafā b. Aḥmad Ḥisārī Akhtarī (d. 968/1560–1561) (fols. 146ᵇ, 147ᵇ, 148ᵇ, 149ᵃ), Fīrūzābādī's *Qāmūs* (fols. 147ᵃ, 147ᵇ, 148ᵃ), Muḥammad b. Abū Bakr Rāzī's (d. 660/1261) *Mukhtār al-Ṣiḥāḥ* by Ismāʿīl Jawharī (d. 1003/1593–1594) (fols. 146ᵇ, 147ᵃ, 148ᵃ, 19ᵃ), Abū l-Suʿūd Efendī's (d. 982/1574) *Tafsīr* (fol. 148ᵃ), Jurjānī's *Taʿrīfāt* (fols. 149ᵃ, 149ᵇ, 152ᵃ, 152ᵇ), Zamakhsharī's *Kashshāf* (fols. 148ᵃ), a certain *Sharḥ Mashāriq* (fol. 152ᵃ), a work by a certain Shaykhzāda (fol. 152ᵃ), Shams al-Dīn al-Maqdasī's (d. 744/1343) commentary on Ibn al-Faraḥ al-Ishbīlī's (d. 671/1273) *Qasīda fī alqāb al-ḥadīth* (fols. 153ᵃ, 153ᵇ, 154ᵃ), and Yaḥyā b. ʿAlī Nawʿī Efendī's (d. 1007/1599) *Natāʾij al-funūn wa-maḥāsin al-mutūn* (in Turkish, fol. 153).

Incipit:

الحمد لله الذي ليس لماهيّته حدّ فيعرب عنه لسان ولم يدخل معرفة ذاته تحت رسم الحدوث والإمكان [...]

Explicit:

[...] والفراسة خاطر يهجم على القلب [...] وقيل مكاشفة اليقين ومعاينة الغيب والله أعلم [بـ]ـحقيقة كلّ شيء.

س = Bagdatli vehbi 2152, held at the Süleymaniye Library, fols. 1–85ᵇ, 13 lines per page. Red-ruled and contains catchwords. The copy, as indicated in the colophon, was completed in 1181 (1767–1768). The main text is preceded by a table of contents executed in *naskhī* hand, but the main text is in *nastaʿlīq*. It is fully pointed and partially vocalized. The chapter and section headings are in red ink, and the entries are marked by a red line above them. Only the first three folios are paginated. The manuscript contains marginal notes copied by a hand similar to the hand of the main text. The notes are quotations from Muḥammad b. Yūsuf Qūnawī Qara-pīrī's *Tarjumān al-Ṣiḥāḥ* (fols. 3ᵃ, 4ᵃ), *al-Akhtarī* by Ḥisārī Akhtarī (fols. 1ᵇ, 2ᵇ, 3ᵃ), Fīrūzābādī's *Qāmūs* (fols. 2ᵇ, 3ᵃ, 5ᵇ), Muḥammad b. Abū Bakr Rāzī's *Mukhtār al-Ṣiḥāḥ* by Ismāʿīl Jawharī (fols. 3ᵃ, 27ᵃ), Abū l-Suʿūd Efendī's *Tafsīr* (fol. 3ᵇ).

Incipit:

الحمد لله الذي ليس لماهيّته حدّ فيعرب عنه لسان ولم يدخل معرفة ذاته تحت رسم الحدوث والإمكان [...]

Explicit:

[...] والفراسة خاطر يهجم على القلب [...] وقيل مكاشفة اليقين ومعاينة الغيب والله أعلم [بـ]حقيقة كلّ شيء.

Colophon:

تمّت سنة 1181 ماه جمادى الآخر

ب = British Library Or. 3143 (88 fols.); 7.75×5.25; 23 lines per page; written in small *nastaʿlīq*; with catchwords. It is fully pointed and contains some marginal notes (one in Persian, fol. 2[a]), which are citations from a certain *Sharḥ al-Tadhkira*, Taftāzānī's *Muṭawwal* and Jurjānī's glosses on it. Only a few corrections appear on the margins. According to Rieu, the manuscript was probably copied in the 12th/18th century. On the title page, the copy is attributed, by a later hand, to Abū l-Faḍl ʿAbd al-Raḥmān Jalāl al-Dīn al-Suyūṭī as follows:

كتاب مقاليد العلوم في الحدود والرسوم تصنيف الشيخ الإمام العالم العامل فريد العصر ووحيد الدهر أبي الفضل عبد الرحمن جلال الدين السيوطي رحمه الله تعالى آمين

Incipit:

الحمد لله الذي ليس لماهيّته حدّ فيعرب عنه لسان ولم يدخل/تدخل معرفة ذاته تحت رسم الحدث والإمكان [...]

Explicit:

[...] والفراسة خاطر يهجم على القلب [...] وقيل مكاشفة اليقين ومعاينة الغيب والله أعلم. تمت

MSS آ and س both contain some marginal notes. All the marginal notes in MS س can also be found in MS آ. However, the latter contains many more marginal notes, some of which are relevant citations from Jurjānī's *Taʿrīfāt*. The others are in Turkish and are taken from Muḥammad b. Yūsuf Qūnawī Qara-pīrī's *Tarjumān al-Ṣiḥāḥ* and a Turkish lexicon called *al-Akhtarī*.[78]

As explored in the critical apparatus, the three manuscripts share textual errors in 80 instances; MS آ and MS س have some 271 common errors, while MS ب and MS س have 103 common errors, and MS آ and MS ب have 97 common errors. Furthermore, MS ب misses out some details which are found in the two other copies. Given the fact that all the marginal notes in MS س are contained in MS آ, but not *vice versa*, and that the texts of the two manuscripts are quite similar, as indicated by a large number of common textual errors, MS آ was likely the exemplar of MS س. Nevertheless, we cannot rule out the possibility that the two manuscripts had been copied from a third manuscript. Moreover, the Ottoman Turkish marginal notes indicate that these two manuscripts were probably produced in the Turkish-speaking environment of the Ottoman Empire, and possibly not far from the libraries, in which they are currently preserved.

MS ب seems to represent an earlier recension of the text. The text of this manuscript misses out some details which are found in the two other copies. A few examples are given in Table 3.

The types and the context of these variations suggest that it is probably not the scribe who is to blame for missing these parts. They are more likely to be the corrections and additions by the author himself. In other words, MS ب arguably represents the earlier recension of the text and MSS آ and س contain its later recension.

As mentioned earlier, Muḥammad Ibrāhīm ʿUbāda edited the *Maqālīd* based on MS ب (MS British Library Or. 3143). There are a number of instances in which ʿUbāda's edition differs from the present edition, partly because of the mistakes and missing parts in the British Library manuscript and partly because of an inaccurate reading of the manuscript. Some examples are presented in Table 4.

In the critical edition, presented here, all the three copies have been used and treated the same with no priority, and none has been considered as the sole base for the edition. If all of them agreed on a textual error, the error has been corrected in the text, while being indicated in the critical apparatus. Some corrections have been made with the help of the relevant sources. We have also modified the text following modern spelling conventions.

78 Moreover, on fol. 48^b of MS آ the date of Shāh Shujāʿ's death is noted in Turkish.

TABLE 3 Examples of missing parts in the text of MS British Library Or. 3143 (the missing parts are distinguished in red)

فإن نظر إلى هذه البضاعة المزجاة بعين الرضا فهو غاية المسؤول ونهاية المرتجى، وإلّا لم يقع من حسن القبول موقعاً، ولم يجد لشرف المثول موضعاً، فذاك من قلّة استعداد العبد.

الدين اعتقاد الشيء على ما كان النحو عليه، حقّاً كان أو باطلاً. قيل: قول إلهي رادع للنفس يقوّمها ويمنعها من الاسترسال فيما طبعت عليه. وقيل: ما التزمه الإنسان لله عزّ وجلّ.

الاستثناء إخراج بعض ما يتناوله اللفظ. المستثنى المتّصل هو المخرج من متعدّد لفظاً أو معنىً.

المنفتحة بخلافها. المستعلية ما يرتفع اللسان بها إلى الحنك. المنخفضة بخلافها.

البسيط العنصري ما كان جزؤه كالكلّ في الاسم والحدّ.

الشكل الفلكي جسم يحيط به اثنى عشر سطحاً مخمّساً متساوية الأضلاع والزوايا.

الرعب امتلاء الجوف عن الخوف.

السخاء إعطاء ما ينبغي لمن ينبغي.

TABLE 4 Some variations between ʿUbāda's edition and the present edition

Present edition	ʿUbāda's edition
[...] زال عنه الطلاقة والابتهاج وآل أمر طلاقته إلى الارتتاج.	[...] زال عنه الطلاقة والإلهاج وآل أمر طلاقته إلى الارتجاج (ص30).
[...] واعتراه عقيب دعواه حمرة الخجل.	[...] واعتراه ضبب، وعراه حمرة الخجل (30).
فمنها [...] ما ثقفتُ عن فرائد محاوراتهم.	فمنها [...] ما نقفتُ عن فرائد محاوراتهم (31).
المأوّل ما يبيّن فيه احتمال الغلط.	المؤوّل ما بيّن فيه احتمال اللفظ (40).
المُسْنَد ما اتّصل سنده من راويه إلى منتهاه.	المُسْنَد ما اتّصل سنده من رواته إلى منتهاه (41)
الفرائض هي قسمة المواريث.	الفرائض هي قسمة الوارثين (55).
الذبيحة حيوان مأكول لا يحلّ ميتة.	الذبيحة حيوان مأكول لأكل منه (59).

TABLE 4 Some variations between ʿUbāda's edition and the present edition (*cont.*)

Present edition	ʿUbāda's edition
الملّة ما بيّنه الله تعالى على لسان رسوله لعباده ذريعة إلى جواره.	الملّة ما بيّنه الله تعالى على لسان رسوله لعباده وديعة إلى جواره (74).
التطبيق إصابة المخبر في بيان الحجّة.	التطبيق إصابة المَحَزّ في بيان الحجّة (93).
الترقّي أن يذكر معنىً ثمّ يردف بما هو أبلغ منه.	التوقّي أن يذكر معنىً يردف بما هو أبلغ منه (97).
الإغراق ما كان الوصف المبالغ فيه ممكناً عقلاً لا عادة.	الإغراق ما كان الوصف المبالغ فيه ممكناً عقلاً له عادة (103).
العقل المستفاد حصول النظريات لها بحيث لا تعزب عنها.	العقل المستفاد حصول النظريات لها بحيث لا تعز عنها (133).
زاوية القطعة ما يحيط بها قوس القطعة ووترها.	القطعة ما يحيط بها قوس القطعة ووترها (148).
السجّل ما يكتب لرسول أو مجمّز أو رحّال بإطلاق نفقته.	السجّل ما يكتب لرسول أو مجهّز أو رحّال بإطلاق نفقته (159).
عينا العود الثقبتان اللتان على وجهه	عينا العود النقبتان اللتان على وجهه (166).
الزايجة صورة مربّعة أو مدوّرة تعمل لمواقع الكواكب في الفلك لينظر فيها عند الحكم لمولد الطفل، أو غيره.	الزابجة صورة مربّعة، أو مدوّرة تعمل لمواقع الكواكب في الفلك ليظر فيها عند الحكم لمولد الطفل، أو غيره (170).
فتح الباب انصراف القمر عن كوكب واتّصاله بآخر بيته في مقابلة بيت الأوّل.	فتح انصراف القمر عن كوكب واتّصاله بآخر بيته في مقابلة بيت الأوّل (173).
المخشّن ما يجعل أجزاء سطح العضو مختلفة الوضع بعد ملاسة طبيعية أو عارضة عن مادّة لزجة.	المخشّن ما يجعل أجزاء سطح العضو مختلفة الوضع بعد ملامسة طبيعية أو عارضة عن مادّة لزجة (178).
اللاذع ما يفرّق اتّصال العضو بقوّة نفّاذة في مواضع لا تحسّ بانفرادها بل بجملتها.	اللاذع ما يفرّق اتّصال العضو بقوّة نفّاذة في مواضع لا تحسّ بأفرادها بل بجملتها (179).
القاشر ما يبلغ من جلائه إلى إخراج الأجزاء الفاسدة.	الفاشر ما يبلغ من جلائه إلى إخراج الأجزاء الفاسدة (180).

TABLE 4 Some variations between ʿUbāda's edition and the present edition (*cont.*)

Present edition	ʿUbāda's edition
المملّس ما يبسط على سطح عضو خشن فتستر خشونته.	المملّس ما يبسط على سطح عضو خشن فتنقشر خشونته (180).
ديابيطس أن يدوم العطش، و كلّما شرب بال.	ديانيطوس أن يدوم العطش، و كلّما شرب بال (195).
العذيوط أن يكون كثير الشبق، رخو المقعد.	العذيوط أن يكون كثير الشبق، لِعوَ المعقد (196).
الذكاء سرعة اقتراح النتائج.	الذكاء سرعة اقتداح النتائج (200).
الوقاحة لجاج النفس في تعاطي القبيح.	الوقاحة إلحاح النفس في تعاطي القبيح (202).
الضلال اعتقاد الباطل حقّاً [...] أو القبيح جميلاً.	الضلال اعتقاد الباطل حقّاً [...] أو القبح جميلاً (202).
الرعب امتلاء الجوف عن الخوف.	الرعب امتلاء الخوف (204).
الورع ملازمة الأعمال الجميلة.	الورع ملازمة الأعمال الجليلة (206).
الرفث فواحش الكلام في النكاح وأوصاف النساء.	الرفث فواحش الكلام لحسب النكاح، وأوصاف النساء (207).
الفتوّة ما يختصّ به الفتى من الفضائل الإنسانية.	الفنون ما يختصّ به الفتى من الفضائل الإنسانية (208).
البرّ السعة في علم الحقّ وفعل الخير.	السير السعة في علم الخلق، وفعل الخير (209).
الذوق نتيجة الكشوفات بورود الواردات.	الذوق نتيجة الكشوفات بورود الواردات.
الشرب يرادفه.	الشرب يرادفه، الذي نتيجة الكشوفات بدوام المواصلات (212–213).
الريّ نتيجة الكشوفات بدوام المواصلات.	
التواضع [...] تذلّل القلوب لعلّام الغيوب.	التواضع [...] بذلك القلوب لعلّام الغيوب (217).

TABLE 4 Some variations between ʿUbāda's edition and the present edition (*cont.*)

Present edition	ʿUbāda's edition
القناعة ترك التشوّق إلى المفقود [...]	القناعة ترك التشوّف إلى المفقود [...] (217)
العبودية [...] وقيل ترك الاختيار فيما يبدو من الأقدار.	العبودية [...] وقيل ترك الاختيار فيما يبدو من الأنوار (219).
اليقين اتّصال البين [وانفصال ما] بين البين.	اليقين اتّصال الهين ما بين البين (219).
التوحيد إفرادك متوحّداً، وهو أن لا يشهد الحقّ إيّاك لك.	التوحيد إفرادك متوحّداً، وهو أن لا يشهد الحقّ إلّا إيّاك لك (ص 220)
الموحّد [...] أن لا يجري عليه ذكر أخطار ما لا حقيقة له عند الله.	الموحّد [...] أن لا يجري عليه ذكر أخطارها لا حقيقة له عند الله (221).
العارف [...] من بذل مجهوده فيما لله.	العارف [...] من بذل محموداً فيما لله (221).

Appendix 1: Other Works in Prose Dedicated to Shāh Shujāʿ or His Viziers

- *Sharḥ al-Fawāʾid al-Ghiyāthiyya* (Arabic) by Muḥammad al-Bukhārī al-Saʿīdī (fl. 760/1359): A commentary on Aḍud al-Dīn Ījī's *al-Fawāʾid al-Ghiyāthiyya*, completed in 760/1359.
  Source: Ḥājjī Khalīfa, *Kashf al-ẓunūn ʿan asāmī al-kutub wa-l-funūn*. 2 vols. (Tehran, 1387/1967), vol. 2, p. 130.
- *Taʿbīr-i sulṭānī* (Persian) by Qāḍī Ismāʿīl b. Niẓām al-Mulk Abarqūhī: on dream interpretation, completed in 763/1361–1362.
  Manuscript: MS Majlis-i Shūrā-yi Islāmī 12528, copied in 958/1551.
- *Mawāhib-i ilāhī* (Persian) by Muʿīn al-Dīn Yazdī: A chronicle about the Muzzafarids, completed in 767/1365.
  Edition (Partial): Muʿīn al-Dīn Yazdī, *Mawāhib-i ilāhī dar tārīkh-i Āl-i Muẓaffar*, ed. Saʿīd Nafīsī, vol. 1. Tehran: Kitābkhāna u Chāpkhāna-yi Iqbāl, 1326 Sh./1947.
- *Kunūz al-wadīʿa min rumūz al-Dharīʿa* (Persian) by Ẓāfir b. Shams al-Dīn Muḥammad: A Persian translation of Rāghib Iṣfahānī's (d. 502/1108) *al-Dharīʿa ilā makārim al-sharīʿa* on moral advice, completed in 768/1367.
  Manuscripts: Majlis-i Shūrā-yi Islāmī 2242, copied in 1069/1658–1659; Marʿashī 12825; Marʿashī 1838.
  Editions: ʿAlī Chulūngar, 'Taṣḥīḥ u taḥshiya-yi *Kunūz al-wadīʿa min rumūz al-Dharīʿa*', PhD Dissertation: University of Isfahan, Isfahan, 1392 Sh./2013; University of Kurdistan, Sanandaj, 1395 Sh./2016–2017; Ghulām-Riḍā Shamsī, 'Muqaddima, taṣḥīḥ u taḥlīl-i kitāb-i *Kunūz al-wadīʿa min rumūz al-Dharīʿa*', PhD Dissertation: University of Kurdistan, Sanandaj, 1395 Sh./2016–2017.
- *Talkhīṣ al-Tadhkira fī l-ṭibb* (Arabic) by Muḥammad b. Isḥāq b. Abī l-ʿAbbās Abarqūhī. Completed in 776/1374–1375, this is a summary of ʿIzz al-Dīn al-Suwaydī's (d. 690/1292) medical work, *al-Tadhkira*.
  Manuscript: University of Tehran 6363, fols. 1^b–60^b (on the margin), copied by Muḥammad Qāsim Ṭabīb b. Raḍī al-Dīn Mālik, completed on 1st Shaʿbān 1144/31 January 1732.
- *Miftāḥ al-saʿāda* (Persian) by Muḥammad b. ʿAbd al-Karīm Nayrīzī, known as Shihāb: On the necessity of appointing a leader (*imām*) and the conditions and manners of rulers, completed in 777/1375.
  Manuscript: Ayasofya 2086 (held at the Süleymaniye Library), holograph, completed in Rabīʿ I 780/June–July 1378.
  Study: Bihrūz Īmānī, 'Durarī az daryā-yi mutūn (4)', *Payām-i Bahāristān* 6 23 (1393 Sh./2014): 349–356.

 | DOI:10.1163/9789004423367_003

- *Sharḥ al-Adwār wa-l-īqāʿ* (Arabic): A commentary on Ṣafī al-Dīn Urmawī's (d. 693/1294) *al-Adwār wa-l-īqāʿ* on music attributed to a certain Mubārakshāh, completed in 777/1375.
  Manuscripts: British Library Or. 2361, fols. 68b–153a; Topkapı Saray, Ahmed III 3458, 68 ff.
  Persian Translation: Mubārakshāh, *Tarjuma-yi sharḥ-i Mubārakshāh-i Bukhārī bar Adwār-i Urmawī dar mūsīqī*, ed. and Persian trans. Sayyid ʿAbdullāh Anwār. Tehran: Farhangistān-i Hunar, 1392 Sh./2013.
- *al-Muḥīṭ al-aʿẓam wa-l-baḥr al-khaḍam fī taʾwīl kitāb Allāh al-ʿazīz al-muḥkam* (Arabic) by Sayyid Ḥaydar Āmulī, an extensive Qurʾānic commentary, completed in 777/1376.
  Source: Āqā Buzurg Ṭihrānī, *al-Dharīʿa ilā taṣānīf al-Shīʿa*, 25 vols. Beirut: Dār al-Aḍwāʾ, 1403–1406/1983–1986. vol. 20, p. 161.
  Manuscript: Gharawiyya 32 (Library of the Shrine of ʿAlī b. Abī Ṭālib); Marʿashī 301, incomplete in the beginning (missing the dedication) and the end.
  Edition: Sayyid Ḥaydar al-Āmulī, *Tafsīr al-Muḥīṭ al-aʿẓam wa-l-baḥr al-khaḍam fī taʾwīl kitāb Allāh al-ʿazīz al-muḥkam*, ed. Sayyid Muḥsin al-Mūsawī al-Tabrīzī, 4 vols. Qom: Muʾassasa-yi Farhangī u Nashr-i Nūrun ʿAlā Nūr, 1380 Sh./2001–2002.
- *Manāhij al-ṭālibīn fī maʿārif al-ṣādiqīn* (Persian) by ʿAlāʾ al-Dīn Hilālī Qazwīnī: A universal history, completed ca. 778/1376.
  Edition: ʿAlāʾ al-Dīn Hilālī Qazwīnī, *Manāhij al-ṭālibīn fī maʿārif al-ṣādiqīn*, edited by Aḥmad Bihnāmī. Tehran: Kitābkhāna, Mūzi u Markaz-i Asnād-i Markaz-i Shūrā-yi Islāmī, 1397 Sh./2018.
- *Nasīm al-Rabīʿ* (Persian) by Muḥammad b. Aḥmad Ṣūfī Samarqandī: A Persian translation of Jār Allāh Zamakhsharī's *Rabīʿ al-abrār*; a collection of stories and traditions on various subjects, completed in 781/1379 and dedicated to Muẓaffar al-Dīn Salghur Shāh, the vizier of Shāh Shujāʿ.
  Manuscript: Haci Selim Aga 902 (held at the Süleymaniye Library), holograph.
  Study: ʿAlī Nawīdī Malāṭī, 'Nasīm al-Rabīʿ-i Ṣūfī-i Samarqandī: Tarjuma-yi qadīm-i Rabīʿ al-abrār-i Samarqandī', *Nāma-yi Bahāristān* 17 (1389 Sh./2010): 167–180.
  Edition: Majīd Manṣūrī, 'Muqaddama u taṣḥīḥ-i *Nasīm al-Rabīʿ*', MA Dissertation, University of Tehran, 1387 Sh./2008–2009.
- *Ḥall mā lā yanḥall* (Arabic) by Iftikhār al-Dīn Dāmghānī: In the introduction to this work, the author explains that in a *majlis* at the court of Shāh Shujāʿ there was a debate between scholars about whether or not Aḍud al-Dīn Ījī's objection to the position of Zamakhsharī on the Qurʾānic verse '*faʾtū bi-sūratin min mithlihi*' (Q, 2:23) was justified. Afterwards, Shāh Shujāʿ ordered Dāmghānī to write a treatise on this matter to clarify it.
  Manuscript: Majlis-i Shūrā-yi Islāmī 1231, fols. 131[a]–133[a], copied by Muḥammad ʿAlī b. Ḥājjī Maḥmūd Nayrīzī.

- *Qānūn al-siyāsa wa-dustūr al-riyāsa* (Arabic) by an anonymous author; on ethics (*tahdhīb al-akhlāq*), economics (*tadbīr al-amwāl*), politics (*taqwīm al-raʿāyā wa-siyāsatihim*).
  Source: Āqā Buzurg Ṭihrānī, *al-Dharīʿa ilā taṣānīf al-Shīʿa*, 25 vols. Beirut: Dār al-Aḍwāʾ, 1403–1406/1983–1986. vol. 17, p. 22.
  Manuscript: Masjid-i Aʿẓam (Qom) 621.
- *Sharḥ al-Risāla al-shāhiyya fī l-akhlāq* (Arabic): An extensive commentary on Ījī's treatise on ethics by an anonymous author. Dedicated to Jalāl al-Dīn Tūrān Shāh (d. 787/1385), the vizier of Shāh Shujāʿ, from ca. 766/1365 onwards.
  Manuscript: University of Tehran 2924 (213 folios).
- *al-Tawḍīḥ fī sharḥ al-Talwīḥ* (Arabic) by an anonymous author: this is a commentary on Fakhr al-Dīn Khujandī's (fl. 702/1303) *al-Talwīḥ ilā asrār al-Tanqīḥ*. The latter work is itself a critical commentary on Ibn Jamīʿ al-Miṣrī's (d. 594/1198) *al-Taṣrīḥ bi-l-maknūn fī tanqīḥ al-Qānūn*.
  Manuscript: Millī-i Malik 4448.
- *Sharḥ Abyāt al-Muwashshaḥ* (Arabic) by Abū l-Makārim Ḍiyāʾ al-Dīn ʿAlī b. Maḥmūd Kirmānī: A super-commentary on Shams al-Dīn al-Khabīṣī's (d. 731/1330–1331) commentary on Ibn Ḥājib's *al-Kāfiya* on Arabic syntax.
  Manuscript: Majlis-i Shūrā-yi Islāmī 508.
- *Ḥilyat al-muftī* (Persian) by Qāḍī Bahāʾ al-Dīn in two chapters.
  Manuscript: UCLA Minasian 32, copied by Saʿd Allāh b. Ghiyāth al-Dīn, completed in 780/1379
- *Sharḥ al-uḥjiyyāt wa-l-lughaziyyāt al-naḥwiyya* (Arabic) by Jamāl al-Dīn Ḥuyayn b. Ḥasan Yazdī.
  Manuscript: Millī-i Malik 5838.

Appendix 2: Sources of the *Maqālīd al-ʿulūm*

Chapter 1: Exegesis (*al-tafsīr*)

Two terms and their definitions in this chapter can be found in Bayḍāwī's *Minhāj al-wuṣūl*. Two others can be found in Baghawī's *Maʿālim al-tanzīl* (also known as *Tafsīr al-Baghawī*). There are also some similarities between the *Maqālīd al-ʿulūm* and Āmulī's Persian work, *Nafāʾis al-funūn*, on the distinction between *tafsīr* and *taʾwīl*. In the case of the similarities with the passage in the *Nafāʾis al-funūn*, it is more likely that the authors of the *Maqālīd and Nafāʾis al-funūn* had a common source which has remained unknown to us.

Minhāj al-wuṣūl/ Maʿālim al-tanzīl/ Nafāʾis al-funūn	*Maqālīd al-ʿulūm*
وأمّا التفسير وهو الكلام في أسباب نزول الآية وشأنها وقصّتها فلا يجوز إلّا بالسماع بعد ثبوته من طريق النقل (معالم التنزيل، ج 1، 46).	التفسير هو الكلام في أسباب نزول الآية وشأنها وقصّتها وأحكامها من طريق النقل.
وأمّا التأويل فهو صرف الآية إلى معنىً محتمل موافق لما قبلها وما بعدها غير مخالف للكتاب والسنّة من طريق الاستنباط (معالم التنزيل، ج 1، 46).	التأويل صرف الكلام إلى معنىً محتمل موافق لما قبلها وما بعدها غير مخالف للكتاب والسنّة على طريق الاستنباط.
در فرق ميان تفسير و تاويل و در اينجا اختلاف بسيار است. جمعى گفتند: تفسير ظاهر قرآن است و تاويل كشف باطن آن. و بعضى ديگر گفتند: تفسير آن است كه به روايت تعلّق داشته باشد و تاويل آنكه به درايت [...] و گروهى گفتند: تفسير محكمات را باشد و تاويل متشابهات را.	قيل: التفسير كشف ظاهر القرآن، والتأويل كشف باطنه. وقيل: التفسير ما يتعلّق بالرواية، والتأويل ما يتعلّق بالدراية. وقيل: التفسير ما يمكن إدراك البشر إلى حقائقه ومعانيه، والتأويل ما يمتنع إدراكه ﴿وَمَا يَعْلَمُ تَأْوِيلَهُ إِلَّا اللَّهُ﴾.

 | DOI:10.1163/9789004423367_004

(*cont.*)

Minhāj al-wuṣūl/ Maʿālim al-tanzīl/ Nafāʾis al-funūn	*Maqālīd al-ʿulūm*
و بعضی دیگر گفتند: هرچه ادراک بشر در معانی و حقایق آن رسد آن را تفسیر خوانند و هرچه نرسد تاویل. و بنا براین فرمود: ﴿وَما يَعْلَمُ تَأْوِيلَهُ إلّا اللّهُ﴾، و نگفت: وما يعلم تفسيره إلّا الله. وطایفهٔ دیگر گفتند: تفسیر آن است که درو خلاف نکردهاند، و تاویل آنکه خلاف کرده باشند. و بعضی دیگر گفتند: تفسیر بیان حقیقت است و تاویل بیان مقصود (نفائس الفنون، ج 1، 369–370).	وقيل: التفسير ما لم يختلف فيه، والتأويل ما اختلف فيه. وقيل: التفسير بيان الحقيقة، والتأويل بيان المقصود. وقيل: التفسير ما لم يرد فيه من السلف إلّا وجه واحد، والتأويل ما ورد فيه وجوه كثيرة. وقيل: التفسير للمحكمات، والتأويل للمتشابهات.
العامّ لفظ يستغرق جميع ما يصلح له بوضع واحد (منهاج الوصول، 121).	العامّ لفظ يستغرق جميع ما يصلح له بوضع واحد.
وهو الواضح بنفسه أو بغيره (منهاج الوصول، 139).	المبيّن الواضح بنفسه أو بغيره.

Chapter 2: Tradition (*al-ḥadīth*)

Some terms and definitions in this chapter can be found in Ṭayyibī's *Khulāṣa fī maʿrifat al-ḥadīth*. Many of these terms and their definitions can be harked back to some earlier works on the subject, including Badr al-Dīn Ibn Jamāʿa's *al-Manhal al-rawī* and Ibn Ṣalāḥ's *ʿUlūm al-ḥadīth* too.

al-Khulāṣa fī maʿrifat al-ḥadīth	*Maqālīd al-ʿulūm*
الصحيح هو ما اتّصل سنده بنقل العدل الضابط عن مثله وسلم من شذوذ وعلّة (35).	الصحيح ما اتّصل سنده بنقل العدل الضابط عن مثله وسلم عن شذوذ وعلّة.

(*cont.*)

Maqālīd al-ʿulūm	*al-Khulāṣa fī maʿrifat al-ḥadīth*
الحسن ما لا يكون في إسناده متّهم. وقيل: ما عرف مخرجه واشتهر رجاله. وقيل: ما اشتهر راويه بالصدق والأمانة، وقصر عن درجة رجال الصحيح حفظاً وإتقاناً بحيث لا يعدّ ما انفرد به منكراً. وقيل: ما لم يخل رجال إسناده عن مستور غير مغفّل في روايته.	ذكر الترمذي أنّه يريد بالحسن أن لا يكون في إسناده متّهم ولا يكون شاذّاً. [...] وقال الخطابي: هو ما عرف مخرجه واشتهر رجاله. [...] وقال ابن صلاح: هو قسمان [...] أحدهما: ما لم يخل رجال إسناده عن مستور غير مغفّل في روايته وقد روي مثله أو نحوه من وجه آخر. والثاني ما اشتهر راويه بالصدق والأمانة، وقصر عن درجة رجال الصحيح حفظاً وإتقاناً بحيث لا يعدّ ما انفرد به منكراً (39).
الضعيف ما لم يجتمع فيه شروط الصحيح والحسن.	في الضعيف وهو كلّ حديث لم يجتمع فيه شروط الصحيح ولا شروط الحسن (48).
المسند ما اتّصل سنده من راويه إلى منتهاه.	قال الخطيب: هو ما اتّصل سنده من راويه إلى منتهاه (39).
المتّصل ما اتّصل رواته، وكان كلّ واحد منهم قد سمعه ممن فوقه، سواء كان مرفوعاً إلى النبيّ، صلّى الله عليه وسلّم، أو موقوفاً على غيره.	والمتّصل ويسمّى أيضاً الموصول وهو كلّ ما اتّصل إسناده، وكان كلّ واحد من رواته قد سمعه ممن فوقه، سواء كان مرفوعاً إلى النبيّ، صلّى الله عليه وسلّم، أو موقوفاً على غيره (50).
المرفوع ما أضيف إلى النبيّ، صلّى الله عليه وسلّم، من قول أو فعل أو تقرير، سواء كان متّصلاً أو منقطعاً.	والمرفوع وهو ما أضيف إلى النبيّ، صلّى الله عليه وسلّم، خاصّة، من قول أو فعل أو تقرير، سواء كان متّصلاً أو منقطعاً، هذا هو المشهور (50).
المعضل ما سقط من سنده اثنان فصاعداً.	المعضل [...] وهو ما سقط من سنده اثنان فصاعداً (75).

(*cont.*)

al-Khulāṣa fī maʿrifat al-ḥadīth	*Maqālīd al-ʿulūm*
الشاذّ والمنكر، قال الشافعي: هو ما رواه الثقة مخالفاً لما رواه الناس. وقال الخليلي: هو ما ليس له إلا إسناد واحد شذّ به شيخ، ثقة كان أو غير ثقة (76–77).	الشاذّ ما رواه الثقة مخالفاً لما رواه الناس. وقيل: ما ليس له إلا إسناد واحد شذّ به شيخ، ثقة كان أو غير ثقة.
الحديث المعلّل هو الذي اطّلع فيه على ما يقدح في صحّته مع أنّ ظاهره السلامة منه (78).	المعلّل ما اطّلع فيه على ما يقدح في صحّته مع أنّ ظاهره السلامة منه.
السند هو الإخبار عن طريق المتن (28).	السند إخبار عن طريق المتن.
الإسناد هو رفع الحديث إلى قائله (28).	الإسناد رفع الحديث إلى قائله.

Chapter 4: Legal Theory (*uṣūl al-fiqh*)

The following terms and definitions can be found in Bayḍāwī's *Minhāj al-wuṣūl*. The sources for the other terms and definitions have remained unknown to us.

Minhāj al-wuṣūl	*Maqālīd al-ʿulūm*
أصول الفقه معرفة دلائل الفقه إجمالاً، وكيفية الاستفادة منها، وحال المستفيد. (51)	أصول الفقه معرفة دلائل الفقه إجمالاً، وكيفية الاستفادة منها، وحال المستفيد.
الحكم خطاب الله تعالى المتعلّق بأفعال المكلّفين بالاقتضاء أو التخيير (53)	الحكم خطاب الله تعالى المتعلّق بأفعال المكلّفين بالاقتضاء أو التخيير.
ويرسم الواجب بأنّه الذي يُذمّ شرعاً تاركه قصداً مطلقاً. ويرادفه الفرض. وقالت الحنفية: الفرض ما ثبت بقطعي والواجب بظنّي (55).	الواجب ما يُذمّ تاركه قصداً مطلقاً. وقيل: ما ثبت وجوبه بدليل غير قطعي. [...] والفرض يرادفه.

(*cont.*)

Minhāj al-wuṣūl	*Maqālīd al-ʿulūm*
المندوب ما يُحمد فاعله ولا يُذمّ تاركه، ويسمّى سنّة ونافلة (56).	الندب ما يحمد فاعله ولا يذمّ تاركه. النافلة والسنّة ترادفانه.
الاشتقاق وهو ردّ لفظ إلى لفظ آخر لموافقته له في حروفه الأصلية، ومناسبته له في المعنى (82).	الاشتقاق ردّ لفظ إلى آخر لموافقته في حروفه الأصلية، ومناسبته في المعنى.
وهو (= الترادف) توالي الألفاظ المفردة الدالّة على معنىً واحد باعتبار واحد (85).	الترادف توالي الألفاظ المفردة الدالّة على شيء واحد باعتبار واحد.

Chapter 5: Rational Theology (*uṣūl al-kalām*)

A number of the terms and definitions in this chapter are taken from Samarqandī's *al-Ṣaḥāʾif al-ilāhiyya*. The table below provides some examples. However, the author must have consulted one other source at least, since the definitions of a number of the terms in the last section of this chapter are not identical with those in *al-Ṣaḥāʾif al-ilāhiyya*.

al-Ṣaḥāʾif al-ilāhiyya	*Maqālīd al-ʿulūm*
إنّه علم يبحث فيه عن ذات الله تعالى وصفاته وأحوال الممكنات في المبدأ والمعاد على قانون الإسلام (66).	علم الكلام ما يبحث فيه عن ذات الله تعالى وصفاته وأحوال الممكنات في المبدأ والمعاد على قانون الإسلام.
كلّ ما يتصوّره العقل [...] إن اقتضت [ذاته] وجوده في الخارج فهو الواجب (67).	الواجب ما اقتضت ذاته وجوده في الخارج.
كلّ ما يتصوّره العقل [...] إن اقتضت [ذاته] عدمه فيه [= في الخارج] فهو الممتنع (67).	الممتنع ما اقتضت ذاته عدم وجوده في الخارج.

(*cont.*)

al-Ṣaḥāʾif al-ilāhiyya	*Maqālīd al-ʿulūm*
كلّ ما يتصوّره العقل [...] إن لم تقتض [ذاته] شيئاً منهما [= من الوجود والعدم] فهو الممكن (67).	الممكن ما لم تقتض ذاته في الخارج وجوداً ولا عدماً.
الوضعي [الوضع؟] هو كون الشيء مشاراً اليه إشارة حسّية (67).	الوضع كون الشيء مشاراً اليه إشارة حسّية.
[...] والحالّ إن كان مبدءاً للآثار المختصّة بالنوع فهو الصورة النوعية (67).	الصورة النوعية حالّة تكون مبدأ الآثار المختصّة بالنوع.
الكيف وهو العرض الذي لا تكون ماهيته بالقياس إلى الغير، ولا يقتضي الانقسام لذاته (69).	الكيف العرض الذي لا تكون ماهيته بالقياس إلى الغير، ولا يقتضي الانقسام لذاته.
حصول الجوهر في الحيّز إن كان بعد حصوله في آخر فهو الحركة. وإن كان بعد حصوله فيه فهو السكون (70).	الحركة حصول الجوهر في حيّز قبل حصوله في حيّز آخر. السكون حصوله بعد حصوله في حيّز آخر.
حصول الجوهرين في حيّزين إن كان بحيث يمكن أن يتوسّطهما ثالث فهو الافتراق، وإلّا فهو الاجتماع (70).	الافتراق حصول جوهرين في حيّزين بحيث يمكن أن يتوسّطهما ثالث. الاجتماع حصولهما فيهما بحيث لا يمكن توسّطه.
[...] فتحقّق أنّ الوجود هو الكون [...] وعلم أيضاً مفهوم العدم إذ هو رفع هذا المفهوم (75–76).	الوجود هو الكون. العدم هو انتفاء الكون.
الوجودي هو ما ليس في نفس مفهومه وحقيقته نفي شيء (96).	الوجودي ما ليس في نفس مفهومه وحقيقته نفي شيء.
الحدوث هو كون الوجود مسبوقاً بالعدم. [...] وقال قوم من المتكلّمين إنّ الحدوث هو الخروج من العدم إلى الوجود (121).	الحدوث كون الوجود مسبوقاً بالعدم. وقيل: هو الخروج من العدم إلى الوجود.
القدم كون الوجود غير مسبوق بالعدم (121).	القدم كون الوجود غير مسبوق بالعدم.

(*cont.*)

al-Ṣaḥāʾif al-ilāhiyya	*Maqālīd al-ʿulūm*
ثمّ الحدوث إمّا إضافي وهو كون ما مضى من وجود شيء أقلّ مما مضى من وجود آخر (121).	الحدوث الإضافي ما مضى من وجود شيء أقلّ مما مضى من وجود آخر.
العلّة ما يحتاج إليه الشيء وهي إمّا تامّة وهي جملة ما يحتاج إليه وجود الشيء، أو غير تامّة وهي خمس؛ لأنّها إمّا جزء لذلك الشيء أو لا. والأوّل علّة الماهية والثاني علّة الوجود. وعلّة الماهية إمّا أن يكون بها الشيء بالقوّة وهي العلّة المادّية، أو بالفعل وهي العلّة الصورية. وعلّة الوجود إمّا أن يكون بها الوجود أو لا، والأوّل العلّة الفاعلية. والثاني إن كان لأجلها الوجود فهي العلّة الغائية [...] (131–132).	العلّة ما يحتاج إليه وجود الشيء. العلّة التامّة جملة ما يحتاج إليه وجود الشيء. العلّة الغير التامّة بعض ما يحتاج إليه وجود الشيء. العلّة المادّية ما يكون بها الشيء بالقوّة. العلّة الصورية ما يكون بها الشيء بالفعل. العلّة الفاعلية ما يفيد وجود الشيء. العلّة الغائية ما يكون لأجلها الشيء.
العلّة الفاعلية وهي ما به وجود الشيء، وهو إمّا مختار وهو الذي إن شاء فعل وإن شاء ترك (133).	الفاعل المختار هو الذي إن شاء فعل وإن شاء ترك.
العالم عند المتكلّمين كلّ ما يوجد سوى الله تعالى (399).	العالم كلّ موجود سوى الله تعالى.
قال المحقّقون: هو تصديق الرسول بكلّ ما علم بالضرورة مجيئه به. [...] ونقل عن السلف أنّه التصديق بالجنان والإقرار باللسان والعمل بالأركان (450).	الإيمان تصديق الرسول بما علم مجيئه به ضرورة. وقيل: إقرار باللسان، وتصديق بالجنان، وعمل بالأركان.

Chapter 6: Dialectic (*al-jadal*)

Some of the terms and definitions in this chapter are taken from Samarqandī's *Risāla fī ādāb al-baḥth*. These terms are presented in the table below. Moreover, the definitions of *ʿilm al-jadal, al-baḥth, al-ḍalāla, al-fahm, al-tafhīm, al-ilhām, al-ṣawāb*, and *al-fikr* can be found in Quṭb al-Dīn Gīlānī's (fl. in or before 826/1423) commentary on Samarqandī's *Risāla*. Most of these terms and definitions (*ʿilm al-jadal, al-baḥth, al-ḍalāla, al-fahm, al-tafhīm, al-ilhām*) are included in Jurjānī's *Taʿrīfāt* too. Apparently, the author of the *Maqālīd*, Gīlānī and Jurjānī used a common source for these definitions, which remained unidentified to us.[1]

Risāla fī ādāb al-baḥth	*Maqālīd al-ʿulūm*
المناظرة هي النظر بالبصيرة من الجانبين في النسبة بين الشيئين إظهاراً للصواب (3).	المناظرة النظر بالبصيرة من الجانبين في النسبة بين الشيئين إظهاراً للصواب.
الأمارة هي التي يلزم بها الظنّ بوجود المدلول (3).	الأمارة ما يلزم من العلم به الظنّ بوجود المدلول.
وما يتوقّف عليه وجود الشيء إن كان داخلاً فيه يسمّى ركناً (3).	أركان الشيء أجزاؤه في الوجود التي لا تحصل إلّا بحصولها داخلة في حقيقته محقّقةً لهويته.
والعلّة التامة جملة ما يتوقّف عليه وجود الشيء (3).	العلّة التامّة جملة ما يتوقّف عليه وجود الشيء.
والتعليل هو تبيين عليّة الشيء (3).	التعليل تبيين علّية الشيء [...]
والملازمة هي كون الحكم مقتضياً للآخر [...] (3).	الملازمة هي كون الحكم مقتضياً لآخر.
والدوران هو ترتّب الشيء على الشيء الذي له صلوح العلّية إمّا وجوداً أو عدماً أو معاً. والأوّل هو الدائر، والثاني هو المدار (3).	الدوران ترتّب الشيء على الشيء الذي له صلوح العلّية وجوداً أو عدماً أو معاً. الدائر هو المرتّب. المدار هو المرتّب عليه.

1 See Quṭb al-Dīn al-Kīlānī, *Sharḥ al-Risāla fī Ādāb al-baḥth*, MS Majlis-i Shūrā-yi Islāmī 10361/6, pp. 68–119 (68–70); al-Jurjānī, *Kitāb al-Taʿīfāt*, pp. 51, 61, 88, 180, and 212.

(*cont.*)

Risāla fī ādāb al-baḥth	*Maqālīd al-ʿulūm*
والمناقضة هي منع مقدّمة الدليل (3).	المناقضة منع مقدّمة معيّنة من مقدّمات دليل الخصم.
والمعارضة هي إقامة الدليل على خلاف ما أقام الدليل عليه الخصم (3).	المعارضة إقامة الدليل على خلاف ما أقام الدليل عليه الخصم.
والنقض هو تخلُّف الحكم عن الدليل (3).	النقض تخلّف الحكم عن الدليل.
والمستند ما يكون المنع مبنيّاً عليه (3).	السند ما يكون المنع مبنياً عليه.

Chapter 7: Syntax (*al-naḥw*)

With the exception of a few instances, all the terms and definitions in this chapter are taken from Ibn Ḥājib's *al-Kāfiya*. The table below provides some examples. The author has also used Tāj al-Dīn Isfarāʾinī's *al-Lubāb fī ʿilm al-iʿrāb* for a definition.

al-Kāfiya/al-Lubāb	*Maqālīd al-ʿulūm*
الكلمة لفظ وضع لمعنىً مفرد (الكافية، 11).	الكلمة [...] وقيل لفظ وضع لمعنىً مفرد.
[...] الكلمة هي اللفظ الموضوع للمعنى مفرداً (اللباب، ص 13)	وقيل: هي اللفظ الموضوع للمعنى مفرداً.
الكلام ما تضمّن كلمتين بالإسناد (الكافية، 11).	الكلام [...] وقيل هو ما تضمّن كلمتين بالإسناد.
الإعراب ما اختلف آخره به ليدلّ على المعاني المعتورة عليه (الكافية، 11).	الإعراب ما اختلف آخره به ليدلّ على المعاني المعتورة عليه.
العامل ما به يتقوّم المعنى المقتضي للإعراب (الكافية، 11).	العامل ما به يتقوّم المعنى المقتضي للإعراب.

(*cont.*)

al-Kāfiya/al-Lubāb	*Maqālīd al-ʿulūm*
فالعدل خروجه عن صيغته الأصلية تحقيقاً كثلاث ومثلث وأخر وجمع أو تقديراً كعمر (الكافية، 12).	العدل خروج الاسم عن صيغته الأصلية تحقيقاً أو تقديراً.
المرفوعات هو [كذا] ما اشتمل على علم الفاعلية (الكافية، 14).	المرفوعات ما اشتمل على علم الفاعلية.
الفاعل وهو ما أسند إليه الفعل أو شبهه وقدّم عليه على جهة قيامه به (الكافية، 14).	الفاعل [...] وقيل: ما أسند إليه الفعل أو شبهه وقدّم عليه على جهة قيامه به.
[مفعول ما لم يسمّ فاعله] كلّ مفعول حذف فاعله وأقيم هو مقامه (الكافية، 15).	مفعول ما لم يسمّ فاعله كلّ مفعول حذف فاعله وأقيم هو مقامه.

Chapter 8: Morphology (*al-ṣarf*)

With a few exceptions, such as the definitions of *afʿāl al-ṭabāʾiʿ*, *al-fiʿl al-lāzim*, *al-wāqiʿ* and *al-mujāwiz*, all the terms and definitions in this chapter are taken from *al-Shāfiya*. At least on one occasion, Jārbardī's commentary on the *Shāfiya* might have been consulted too. The table below provides some examples.

al-Shāfiya/Sharḥ al-Shāfiya	*Maqālīd al-ʿulūm*
التصريف علم بأصول يتعرّف بها أحوال أبنية الكلم التي ليست بإعراب (الشافية، 59).	علم الصرف أصول يتعرّف بها أحوال أبنية الكلم التي ليست بإعراب.
المعتلّ ما فيه حرف علّة (الشافية، 60).	المعتلّ ما فيه حرف علّة.
الآلة هي كلّ اسم اشتقّ من فعل اسماً يستعان به في ذلك الفعل (شرح الشافية، ج 1، 320).	الآلة كلّ اسم اشتقّ من فعل اسماً لما يستعان به في ذلك الفعل.

(*cont.*)

al-Shāfiya/Sharḥ al-Shāfiya	*Maqālīd al-ʿulūm*
المنسوب الملحق آخره ياء مشدّدة لتدلّ على نسبته إلى المجرّد عنها (الشافية، 70).	المنسوب الملحق بآخره ياء مشدّدة ليدلّ على نسبته إلى المجرّد عنها.
[...] والروم في المتحرّك وهو أن تأتي بالحركة خفيّة (الشافية، 79).	الروم أن تأتي بالحركة خفيّة.
والإشمام في المضموم وهو أن تضمّ الشفتين بعد الإسكان (الشافية، 79).	الإشمام أن تضمّ الشفتين بعد الإسكان.
المقصور ما آخره ألف مفردة (الشافية، 80).	المقصور ما آخره ألف مفردة.
الممدود ما يكون آخره ألف بعدها همزة (الشافية، 80).	الممدود ما يكون آخره ألف بعدها همزة.
الإمالة أن ينحى بالفتحة نحو الكسرة وسببها قصد المناسبة لكسرة أو ياء (الشافية، 85)	الإمالة أن تنحى بالفتحة نحو الكسرة لقصد المناسبة.
الإعلال تغيير حرف العلّة للتخفيف (الشافية، 88).	الإعلال تغيير حروف العلّة للتخفيف.
الإدغام أن تأتي بحرفين ساكن فتحرّك من مخرج واحد من غير فصل (الشافية، 95)	الإدغام أن تأتي بحرفين ساكن فتحرّك من مخرج واحد بلا فصل.
المجهورة ما ينحصر جري النفس مع تحرّكه (الشافية، 97)	الحروف المجهورة ما ينحصر جري النفس مع تحرّكه.
الشديدة ما ينحصر جري صوته عند إسكانه في مخرجه (الشافية، 97)	الشديدة ما ينحصر جري صوته عند إسكانه في مخرجه.

Chapter 9: Rhetoric (*al-maʿānī wa-l-bayān*)

Not all the sources for this chapter are known to us. A major source, however, for a number of the definitions is Khaṭīb Qazwīnī's *al-Īḍāḥ fī ʿulūm al-balāgha*. Furthermore, Sakkākī's *Miftāḥ al-ʿulūm* must have been consulted by the author. A few of the definitions can be found in both Sakkākī's *Miftāḥ* and Qazwīnī's *al-Īḍāḥ*. Ṭayyibī's *al-Tibyān fī l-bayān* may have been the source for a few terms and definitions too. Moreover, the definition of one term (*al-faṣāḥa*) can be found in Ibn Nāẓim's *al-Miṣbāḥ*.

Miftāḥ al-ʿulūm/al-Īḍāḥ/al-Tibyān/al-Miṣbāḥ	*Maqālīd al-ʿulūm*
علم المعاني هو تتبّع خواصّ تراكيب الكلام في الإفادة وما يتّصل بها من الاستحسان وغيره، ليحترز بالوقوف عليها عن الخطأ في تطبيق الكلام على ما يقتضي الحال ذكره (مفتاح العلوم، 161؛ الإيضاح، 23).	علم المعاني تتبّع خواصّ تراكيب الكلام في الإفادة وما يتّصل بها من الاستحسان وغيره، ليحترز بالوقوف عليها عن الخطأ في تطبيق الكلام على ما يقتضي الحال ذكره.
أمّا بلاغة الكلام فهي مطابقته لمقتضى الحال مع فصاحته (الإيضاح، 20).	البلاغة في الكلام مطابقته لمقتضى الحال مع فصاحته.
كالأسلوب الحكيم فيها وهو تلقّي المخاطب بغير ما يترقّب [...] أو السائل بغير ما يتطلّب. (مفتاح العلوم، ص 327)	الأسلوب الحكيم تلقّي المخاطب بغير ما يترقّب وتطلّب السائل بغير ما يتطلّب.
تعريف البلاغة هي بلوغ المتكلّم في تأدية المعنى حدّاً له اختصاص بتوفية خواصّ التراكيب حقّها، وإيراد أنواع التشبيه والمجاز والكناية على وجهها (مفتاح العلوم، 415؛ الإيضاح، 24).	بلاغة المتكلّم بلوغه في تأدية المعنى حدّاً له اختصاص بتوفية خواصّ التراكيب حقّها، وإيراد أنواع التشبيه والمجاز والكناية على وجهها.
وأمّا بلاغة المتكلّم فهي ملكة يقتدر بها على تأليف كلام بليغ (الإيضاح، 21).	بلاغة المتكلّم [...] وقيل ملكة يقتدر بها على تأليف كلام بليغ.

(*cont.*)

Miftāḥ al-ʿulūm/al-Īḍāḥ/al-Tibyān/al-Miṣbāḥ	*Maqālīd al-ʿulūm*
الفصاحة هي صوغ الكلام على وجه له توفية بتمام الإفهام لمعناه وتبيين المراد منه (المصباح، 59).	الفصاحة صوغ الكلام على وجه له توفية بتمام الإفهام لمعناه وتبيين المراد وتزيين الألفاظ بما يقرب فهمه، ويعزب نظمه [...]
مراعاة النظير [...] هي أن يجمع في الكلام بين أمر وما يناسبه لا بالتضادّ (الإيضاح، 260).	مراعاة النظير جمع أمر وما يناسبه لا بالتضادّ.
التغليب وهو ترجيح أحد المعلومين على الآخر وإطلاق لفظه عليهما. (التبيان، 136).	التغليب ترجيح أحد المعلومين على الآخر وإطلاق لفظه عليهما.
[...] ابتدائي وهو ما خوطب به خالي الذهن (التبيان، 9).	الكلام الابتدائي ما خوطب به خالي الذهن.
[...] إنكاري وهو ما ردّ به حكم المخالف (التبيان، 9).	الكلام الإنكاري ما ردّ به حكم المخالف.
الجامع العقلي هو أن يكون بينهما اتّحاد في التصوّر (مفتاح العلوم، 253).	الجامع العقلي أن يكون بين الجملتين اتّحاد في التصوّر.
الوهمي هو أن يكون بين تصوّراتهما شبه تماثل (مفتاح العلوم، 253–254).	الجامع الوهمي أن يكون بين تصوّراتهما شبه تماثل.
الخيالي هو أن يكون بين تصوّراتهما تقارن في الخيال سابق لأسباب مؤدّية إلى ذلك (مفتاح العلوم، 254).	الجامع الخيالي أن يكون بين تصوّراتهما تقارن في الخيال سابق لأسباب مؤدّية إلى ذلك.
الإيجاز هو أداء المقصود من الكلام بأقلّ من عبارات متعارف الأوساط، والإطناب أداؤه بأكثر من عباراتهم (مفتاح العلوم، 277؛ الإيضاح، 139).	الإيجاز أداء المقصود بأقلّ من عبارة المتعارف. الإطناب أداؤه بأكثر منها.
ومنه التفريق وهو إيقاع تباين بين أمرين من نوع واحد [...]	التفريق إيقاع تباين بين أمرين من نوع واحد.

(*cont.*)

Miftāḥ al-ʿulūm/al-Īḍāḥ/al-Tibyān/al-Miṣbāḥ	*Maqālīd al-ʿulūm*
ومنه التقسيم وهو ذكر متعدّد ثمّ إضافة ما لكلّ إليه على التعيين [...] ومنه الجمع مع التفريق وهو أن يدخل شيئان في معنى واحد ويفرّق بين جهتي الإدخال [...] ومنه الجمع مع التقسيم وهو جمع متعدّد تحت حكم ثمّ تقسيمه، أو تقسيمه ثمّ جمعه. (الإيضاح، 269–271).	التقسيم ذكر متعدّد ثمّ إضافة ما لكلّ إليه على التعيين. الجمع مع التفريق أن يدخل شيئان في معنىً ويفرّق بين جهتي الإدخال. الجمع مع التقسيم أن يجمع متعدّد تحت حكم ثمّ يقسّم أو بالعكس.
لزوم ما لا يلزم وهو أن يجئ [قبل] حرف الرويّ وما في معناه من الفاصلة ما ليس بلازم في مذهب السجع (الإيضاح، 300).	لزوم ما لا يلزم أن يجئ [قبل] حرف الرويّ أو ما في معناه من الفاصلة ما ليس بلازم في السجع.
ومنه التشريع وهو بناء البيت على قافيتين يصحّ المعنى على الوقوف على كلّ واحدة منهما (الإيضاح، 300).	التشريع بناء البيت على قافيتين يصحّ المعنى على الوقوف على كلّ منهما.
الموازنة: تساوي الفاصلتين في الوزن دون التقفية (الإيضاح، 300).	الموازنة تساوي الفاصلتين في الوزن دون التقفية.
ومنه السجع وهو تواطؤ الفاصلتين من النثر على حرف واحد [...]	السجع تواطؤ الفاصلتين من النثر على حرف واحد.
إن اختلفا في الوزن فهو السجع المطرّف [...]	السجع المطرّف اختلاف الفاصلتين في الوزن.

(*cont.*)

Miftāḥ al-ʿulūm/al-Īḍāḥ/al-Tibyān/al-Miṣbāḥ	*Maqālīd al-ʿulūm*
وإلّا فإن كان ما في إحدى القرينتين من الألفاظ أو أكثر ما فيها مثل ما يقابله من الأخرى في الوزن والتقفية فهو الترصيع [...] وإلّا فهو السجع المتوازي (الإيضاح، 296).	السجع المتوازي ما فيه إحدى القرينتين أو أكثر مثل ما يقابله من الأخرى.

Chapter 10: Prosody (*al-ʿarūḍ*)

The major source for this chapter is Tabrīzī's *al-Kāfī fī ʿilmay al-ʿarūḍ wa-l-qawāfī*. In addition, the author has consulted Khʷārazmī's *Mafātīḥ al-ʿulūm* and Ibn Sarrāj's *al-Kāfī fī ʿilm al-qawāfī*.

al-Kāfī fī ʿilmay al-ʿarūḍ wa-l-qawāfī/ Mafātīḥ al-ʿulūm/al-Kāfī fī ʿilm al-qawāfī	*Maqālīd al-ʿulūm*
الابتداء اسم لكلّ جزء يعتلّ في أوّل البيت بعلّة لا تكون في شيء من الحشو كالخرم (الكافي في علمي العروض والقوافي، 141).	الابتداء اسم لكلّ جزء معتلّ في أوّل البيت بعلّة لا تكون في شيء من الحروف كالخرم.
الاعتماد اسم للأسباب التي تزاحفها (الكافي في علمي العروض والقوافي، 141).	الاعتماد اسم للأسباب التي تزاحفها.
السبب الثقيل حرفان متحرّكان [...] الوتد المجموع ثلاثة أحرف، الأوّل والثاني متحرّكان والثالث ساكن [...] الوتد المفروق ثلاثة أحرف، الأوّل والثالث متحرّكان وبينهما ساكن [...]	السبب الخفيف حرفان أوّلهما متحرّك وثانيهما ساكن. السبب الثقيل حرفان متحرّكان. الوتد المجموع ثلاثة أحرف تحرّك منها اثنان وسكن الثالث.

(*cont.*)

al-Kāfī fī ʿilmay al-ʿarūḍ wa-l-qawāfī/ Mafātīḥ al-ʿulūm/al-Kāfī fī ʿilm al-qawāfī	*Maqālīd al-ʿulūm*
الفاصلة الصغرى أربعة أحرف، ثلاثة منها متحرّكة والرابع ساكن [...]	الوتد المفروق ثلاثة أحرف تحرّك أوّلها وآخرها وسكن أوسطها.
الفاصلة الكبرى خمسة أحرف أربعة منها متحرّكة والخامس ساكن (مفاتيح العلوم، 79–80).	الفاصلة الصغرى أربعة أحرف تحرّكت الثلاثة الأُول منها وسكن الرابع.
	الفاصلة الكبرى خمسة أحرف تحرّكت الأربعة الأُول وسكن الخامس.
المقبوض ما سقط خامسه الساكن.	المقبوض ما أسقط خامسه الساكن.
المكفوف ما سقط سابعه الساكن.	المكفوف ما أسقط سابعه الساكن.
المعاقبة بين الحرفين أن لا يجوز سقوطهما معاً، وإن جاز ثبوتهما معاً.	المعاقبة بين الحرفين أن لا يجوز سقوطهما معاً، وإن جاز ثبوتهما معاً.
الخرم حذف أوّل متحرّك من الوتد المجموع في أوّل البيت.	المراقبة بين الحرفين أن لا يجوز سقوطهما جميعاً، ولا ثبوتهما جميعاً.
الخزم زيادة في أوّل البيت لا يعتدّ بها في التقطيع.	الخرم حذف أوّل متحرّك من الوتد المجموع في أوّل البيت.
الأثلم فعولن إذا خرم.	الخزم زيادة في أوّل البيت لا يعتدّ بها في التقطيع.
السالم ما سلم من الزِحاف.	الأثلم فعولن السالم خاصّة إذا خرم.
المحذوف ما سقط من آخره سبب.	السالم ما سلم من الزِحاف.
المجزوء ما سقط منه جزءان.	المحذوف ما سقط من آخره سبب خفيف.
المخبون ما سقط ثانيه الساكن.	المخروم ما سقط منه جزءان.
المشكول ما سقط ثانيه وسابعه الساكنان.	المخبون ما سقط منه الساكن.
الصدر ما زوحف لمعاقبة ما قبله.	المشكول ما سقط ثانيه وسابعه الساكنان.
العجز ما زوحف لمعاقبة ما بعده.	الصدر ما زوحف لمعاقبة ما قبله فقط.

(*cont.*)

al-Kāfī fī ʿilmay al-ʿarūḍ wa-l-qawāfī/ Mafātīḥ al-ʿulūm/al-Kāfī fī ʿilm al-qawāfī	*Maqālīd al-ʿulūm*
الطرفان ما زوحف لمعاقبة ما قبله وما بعده.	العجز ما زوحف لمعاقبة ما بعده فقط.
البريء ما سلم من هذه المعاقبة.	الطرفان ما زوحف لمعاقبة ما قبله وما بعده.
المقصور ما سقط ساكن سببه وسكن متحركه.	البريء ما سلم من هذه المعاقبة.
المقطوع ما سقط ساكن وتده وسكن متحركه.	المقصور ما سقط ساكن سببه وسكن متحركه.
المطويّ ما سقط رابعه الساكن.	المقطوع ما حذف ساكن وتده وأسكن متحركه.
المخبول ما سقط ثانيه ورابعه الساكنان.	المطويّ ما سقط رابعه الساكن.
المذال ما زيد على اعتداله من عند وتده حرف ساكن.	المخبول ما سقط ثانيه ورابعه الساكنان.
المعصوب ما سكن خامسه [...]	المذال ما زيد على اعتداله من عند وتده حرف ساكن.
المعقول ما سقط خامسه بعد سكونه [...]	المعصوب ما سكن خامسه.
المنقوص ما سقط سابعه بعد سكون خامسه [...]	المعقول ما أسقط خامسه بعد سكونه.
الأعضب خرم مفاعلتن حتّى يصير مفتعلن.	المنقوص ما أسقط سابعه بعد سكون خامسه.
الأقصم خرم مفاعيلن من الوافر حتّى يصير مفعولن.	
الأعقص خرم مفاعيل حتّى يصير مفعول.	العضب خرم مفاعلتن حتّى يصير مفتعلن.
الأجمّ خرم مفاعلن حتّى يصير فاعلن.	الأقصم مفاعلتن متى خرم بعد العضب من الوافر حتّى يصير مفعولن.
المقطوف ما سقط منه زنة سبب خفيف بعد سكون خامسه.	العقص خرم مفاعيل من الوافر حتّى يصير مفعول.
المضمر ما سكن ثانيه.	المقطوف ما سقط منه زنة سبب خفيف بعد سكون خامسه.
[...]	
الأحذّ ما سقط من آخره وتد مجموع.	المضمر ما سكن ثانيه.
المرفّل ما زيد على اعتداله سبب خفيف.	الأحذّ ما أسقط منه الوتد المجموع.
	المرفّل ما زيد على اعتداله سبب خفيف.

(*cont.*)

al-Kāfī fī ʿilmay al-ʿarūḍ wa-l-qawāfī/ Mafātīḥ al-ʿulūm/al-Kāfī fī ʿilm al-qawāfī	***Maqālīd al-ʿulūm***
	الأجمّ مفاعلن في الوافر إذا خرم حتّى يصير فاعلن.
الأخرم خرم مفاعيلن من الهزج حتّى يصير مفعولن.	الأخرم مفاعيلن إذا خرم حتّى يصير مفعولن.
الأخرب خرم مفاعيلن حتّى يصير مفعول.	الأخرب مفاعيلن إذا خرم حتّى يصير مفعول.
الأشتر خرم مفاعلن حتّى يصير فاعلن.	الأشتر مفاعلن إذا خرم حتّى يصير فاعلن.
المشطور ما سقط منه شطره.	المشطور ما أسقط منه شطره.
المنهوك ما أسقط ثلثاه.	المنهوك ما أسقط ثلثاه.
المسبّغ ما زيد على اعتداله من عند سببه حرف ساكن.	المسبّغ ما زيد على اعتداله من عند سببه حرف ساكن.
المكشوف ما حذف متحرّك وتده المفروق.	المكشوف ما حذف متحرك وتده المفروق.
الموقوف ما سكن متحرّك وتده المفروق.	الموقوف ما سكن متحرك وتده المفروق.
الأصلم ما سقط وتده المفروق.	الأصلم ما سقط وتده المفروق.
المشعّث ما سقط أحد متحركي وتده، ولا يكون إلّا في الخفيف والمجتثّ.	المشعّث ما سقط أحد متحركي وتده.
المراقبة بين الحرفين أن لا يجوز سقوطهما جميعاً، ولا ثبوتهما جميعاً.	
الأبتر ما سقط ساكن وتده وسكن متحرّكه وقد سقط من آخره سبب [...] (الكافي في علمي العروض والقوافي، 143–145).	الأبتر ما سقط [ساكن] وتده وسكن متحركه وقد سقط من آخره سبب.
القافية الكلمة الأخيرة من البيت (مفاتيح العلوم، 90).	القافية الكلمة الأخيرة من البيت.
الرويّ الحرف الذي تبنى عليه القصيدة من القافية (مفاتيح العلوم، 90).	حرف الرويّ ما يبنى الشاعر قصيدته عليه.

(*cont.*)

al-Kāfī fī ʿilmay al-ʿarūḍ wa-l-qawāfī/ Mafātīḥ al-ʿulūm/al-Kāfī fī ʿilm al-qawāfī	*Maqālīd al-ʿulūm*
المتكاوس وهو ما كان في آخره فاصلة كبرى (الكافي في علم القوافي، 35).	المتكاوس ما في آخره فاصلة كبرى.
المتراكب وهو ما كان في آخره فاصلة صغرى (الكافي في علم القوافي، 36).	المتراكب ما في آخره فاصلة صغرى.
المترادف وهو ما اجتمع في آخره ساكنان (الكافي في علم القوافي، 37).	المترادف ما يجتمع في آخره حرفان ساكنان.
المتدارك وهو ما كان في آخره وتد مجموع (الكافي في علم القوافي، 36).	المتدارك ما في آخره وتد مجموع.
المتواتر وهو ما كان في آخره سبب خفيف (الكافي في علم القوافي، 37).	المتواتر ما في آخره سبب خفيف.
الإقواء اختلاف حركة الرويّ في قصيدة واحدة [...] الإصراف إقواء بالنصب (الكافي في علمي العروض والقوافي، 160).	الإقواء اختلاف حركة الرويّ في قصيدة واحدة. الإصراف كذلك.
الإكفاء اختلاف حرف الرويّ في قصيدة واحدة. (الكافي في علمي العروض والقوافي، 161).	الإكفاء اختلاف حرف الرويّ في قصيدة واحدة.
الإيطاء أن تتكرّر القافية في قصيدة واحدة بمعنىً واحد (الكافي في علمي العروض والقوافي، 162).	الإيطاء أن تتكرّر القافية في قصيدة واحدة بمعنىً واحد.
التضمين هو أن تتعلّق قافية البيت الأوّل بالبيت الثاني. (الكافي في علمي العروض والقوافي، 166).	التضمين أن تتعلّق قافية البيت الأوّل بالبيت الثاني.

Chapter 11: Logic (*al-manṭiq*)

The major source of this chapter is Kātibī's *Shamsiyya*. The author also adopted a few definitions from Quṭb al-Dīn Rāzī's commentary on the *Shamsiyya*, titled *Taḥrīr al-qawāʾid al-manṭiqiyya*. The table below provides some examples.

al-Shamsiyya* and *Taḥrīr al-qawāʾid al-manṭiqiyya	***Maqālīd al-ʿulūm***
ورسّموه بأنّه [= علم المنطق] آلة قانونية تعصم مراعاتها تعصم مراعاتها الذهن عن الخطأ في الفكر (الشمسية، 52).	علم المنطق آلة قانونية تعصم مراعاتها تعصم مراعاتها الذهن عن الخطأ في الفكر
الفكر ترتيب أمور معلومة للتأدّي إلى مجهول (الشمسية، 52).	الفكر ترتيب أمور معلومة للتأدّي إلى مجهول
القانون أمر كلّي ينطبق على جميع جزئياته ليتعرّف أحكامها منه (تحرير القواعد، 60).	القانون أمر كلّي منطبق على جزئيات يتعرّف أحكامها منه.
العلم إمّا تصوّر فقط وهو حصول صورة الشيء في العقل (الشمسية، 30).	التصوّر حصول صورة الشيء في العقل.
العلم [...] وإما تصوّر معه حكم [...] ويقال للمجموع التصديق (الشمسية، 30).	التصديق تصوّر معه حكم.
[...] حكم وهو إسناد أمر إلى آخر إيجاباً أو سلباً (الشمسية، 30).	الحكم إسناد أمر إلى آخر إيجاباً أو سلباً.
الإيجاب وهو إيقاع النسبة. والسلب هو انتزاعها (تحرير القواعد، 33).	الإيجاب إيقاع النسبة. السلب انتزاعها.
[...] الدلالة وهي كون الشيء بحالة يلزم من العلم به العلم بشيء آخر (تحرير القواعد، 83).	الدلالة كون الشيء بحالة يلزم من العلم به العلم بشيء آخر.

(*cont.*)

al-Shamsiyya* and *Taḥrīr al-qawāʾid al-manṭiqiyya	***Maqālīd al-ʿulūm***
الوضع جعل اللفظ بإزاء المعنى أوّلاً (تحرير القواعد، 84).	الوضع جعل اللفظ بإزاء المعنى أوّلاً.
[...] اللزوم الخارجي وهو كون الأمر الخارجي بحيث يلزم من تحقق المسمى في الخارج تحقّقه في الخارج كما أنّ اللزوم الذهني هو كون الأمر الخارجي بحيث يلزم من تحقّق المسمّى في الذهن تحقّقه في الذهن (تحرير القواعد، 90–91).	اللزوم الذهني كون الأمر الخارجي بحيث يلزم من تحقّق المسمّى في الذهن تحقّقه في الذهن. اللزوم الخارجي كونه بحيث يلزم من تحقّق المسمّى في الخارج تحقّقه فيه.
وهو [= اللازم] إمّا بيّن وهو الذي يكون تصوّره مع تصوّر ملزومه كافياً في جزم الذهن باللزوم بينهما (الشمسية، 154).	اللازم البيّن ما يكون تصوّره مع تصوّر ملزومه كافياً في جزم الذهن باللزوم بينهما.
وإمّا [اللازم] غير بيّن وهو الذي يفتقر جزم الذهن باللزوم بينهما إلى وسط (الشمسية، 154).	اللازم الغير البيّن ما يفتقر جزم الذهن باللزوم بينهما إلى وسط.
ترسم الخاصّة بأنّها كلّية مقولة على ما تحت حقيقة واحدة فقط قولاً عرضياً (الشمسية، 160).	الخاصّة كلّية مقولة على ما تحت أفراد واحدة فقط قولاً عرضياً.
[ويرسم] العرض العامّ بأنّه كلّي مقول على أفراد حقيقة واحدة وغيرها قولاً عرضياً (الشمسية، 160).	العرض العامّ كلّي مقول على أفراد حقيقة وغيرها قولاً عرضياً.
القضية قول يصحّ أن يقال لقائله "إنّه صادق فيه أو كاذب"(الشمسية، 220).	القضية قول يصحّ أن يقال لقائله "إنّه صادق أو كاذب".
وهي [= القضية] حملية إن انحلّت بطرفيها إلى مفردين [...] وشرطية إن لم تنحلّ (الشمسية، 220).	الحملية قضية تنحلّ بطرفيها إلى مفردين. الشرطية ما لا تنحلّ بطرفيها إلى مفردين.

(*cont.*)

al-Shamsiyya* and *Taḥrīr al-qawāʾid al-manṭiqiyya	***Maqālīd al-ʿulūm***
وهي [= المنفصلة] التي يحكم فيها بالتنافي بين القضيتين في الصدق والكذب معاً أو في أحدهما فقط أن بنفيه (الشمسية، 226).	المنفصلة شرطية يحكم فيها بالتنافي بين قضيتين في الصدق والكذب معاً أو في أحدهما فقط.
وموضوع الحملية إن كان شخصاً معيّناً سمّيت مخصوصة وشخصية (الشمسية، 236).	المخصوصة ما يكون الموضوع فيها شخصاً معيّناً. الشخصية مرادفها.
واللفظ الدالّ عليها أي على كمّية الأفراد يسمّى سوراً (تحرير القواعد، 237).	السور اللفظ الدالّ على بيان كمّية أفراد ما عليه الحكم.
وإن لم يبيّن فيها كمّية الأفراد، فإن لم تصلح لأن تصدق كلّية وجزئية سمّيت القضية طبيعية. وإن صلحت لذلک سمّيت مهملة (الشمسية، 241).	الطبيعية ما لم يبيّن فيها كمّية أفراد ما عليه الحكم، ولم يصلح أن تصدق كلّية وجزئية. المهملة ما لم يبيّن فيها كمّية الأفراد، وتكون صالحة للكلّية والجزئية.
حرف السلب إن كان جزءاً من الموضوع [...] أو من المحمول [...] أو منهما جميعاً سمّيت القضية معدولة. [...] وإن لم يكن جزءاً لشيء منهما سمّيت محصّلة إن كانت موجبة، وبسيطة إن كانت سالبة (الشمسية، 263).	المعدولة ما يكون حرف السلب فيها جزءاً من الموضوع أو المحمول أو منهما جميعاً. المحصّلة موجبة لم يكن حرف السلب فيها جزءاً لا من الموضوع، ولا من المحمول. البسيطة سالبة لم يكن حرف السلب فيها جزءاً من الموضوع والمحمول.
[منها] بسيطة وهي التي حقيقتها إيجاب فقط أو سلب فقط (الشمسية، 276).	الموجّهة البسيطة قضية حقيقتها إيجاب فقط أو سلب فقط.
ومنها مركّبة وهي التي حقيقتها تركّبت من إيجاب وسلب معاً (الشمسية، 276).	الموجّهة المركّبة ما تركّبت حقيقتها من إيجاب وسلب.

(*cont.*)

al-Shamsiyya and *Taḥrīr al-qawāʾid al-manṭiqiyya*	*Maqālīd al-ʿulūm*
الضرورية المطلقة وهي التي يحكم فيها بضرورة ثبوت المحمول للموضوع أو سلبه عنه ما دامت ذات الموضوع موجودة (الشمسية، 276).	الضرورية المطلقة ما يحكم فيها بضرورة ثبوت المحمول للموضوع أو سلبه عنه، ما دام ذات الموضوع موجوداً.
الدائمة المطلقة وهي التي يحكم فيها بدوام ثبوت المحمول للموضوع أو سلبه عنه ما دام ذات الموضوع موجودة (الشمسية، 276).	الدائمة المطلقة ما يحكم فيها بدوام ثبوت المحمول للموضوع، ما دام ذات الموضوع موجوداً.
المشروطة العامّة وهي التي يحكم فيها بضرورة ثبوت المحمول للموضوع أو سلبه عنه بشرط وصف الموضوع (الشمسية، 276).	المشروطة العامّة ما يحكم فيها بضرورة ثبوت المحمول للموضوع أو سلبه عنه بشرط وصف الموضوع.
العرفية العامّة وهي التي يحكم فيها بدوام ثبوت المحمول للموضوع أو سلبه عنه بشرط وصف الموضوع (الشمسية، 276).	العرفية العامّة ما يحكم فيها بدوام ثبوت المحمول للموضوع أو سلبه عنه بشرط وصف الموضوع.
المطلقة العامّة وهي التي يحكم فيها بثبوت المحمول للموضوع أو سلبه عنه بالفعل (الشمسية، 276).	المطلقة العامّة ما يحكم فيها بثبوت المحمول للموضوع أو سلبه عنه بالفعل.
الممكنة العامّة وهي التي يحكم فيها بارتفاع الضرورة المطلقة عن جانب المخالف للحكم (الشمسية، 276).	الممكنة العامّة ما يحكم فيها بارتفاع الضرورة المطلقة عن جانب المخالف.
المشروطة الخاصّة وهي المشروطة العامّة مع قيد اللادوام بحسب الذات (الشمسية، 285).	المشروطة الخاصّة ما يحكم فيها بضرورة ثبوت المحمول للموضوع أو سلبه عنه، ما دام ذات الموضوع موجوداً مقيّداً بقيد اللادوام.

(*cont.*)

al-Shamsiyya and *Taḥrīr al-qawāʾid al-manṭiqiyya*	*Maqālīd al-ʿulūm*
العرفية الخاصّة وهي العرفية العامّة مع قيد اللادوام بحسب الذات.	العرفية الخاصّة ما يحكم فيها بدوام ثبوت المحمول للموضوع أو سلبه عنه بشرط وصف الموضوع مع قيد اللادوام.
الوجودية اللاضرورية وهي المطلقة العامّة مع قيد اللاضرورة بحسب الذات (الشمسية، 289).	الوجودية اللاضرورية ما يحكم فيها بثبوت المحمول للموضوع أو سلبه عنه بالفعل مع قيد اللاضرورة.
الوجودية اللادائمة وهي المطلقة العامّة مع قيد اللادوام بحسب الذات (الشمسية، 290–291).	الوجودية اللادائمة ما يحكم فيها بثبوت المحمول للموضوع أو سلبه عنه بالفعل مع قيد اللادوام بحسب الذات.
الوقتية وهي التي يحكم فيها بضرورة ثبوت المحمول للموضوع أو سلبه عنه في وقت معيّن من أوقات وجود الموضوع مع قيد اللادوام بحسب الذات (الشمسية، 291–292).	الوقتية ما يحكم فيها بضرورة ثبوت المحمول للموضوع أو سلبه عنه في وقت معيّن من أوقات وجود الموضوع مقيّداً باللادوام بحسب الذات.
المنتشرة وهي التي حكم فيها بضرورة ثبوت المحمول للموضوع أو سلبه عنه في وقت غير معيّن من أوقات وجود الموضوع مقيّداً باللادوام بحسب الذات (الشمسية، 294).	المنتشرة ما يحكم فيها بضرورة ثبوت المحمول للموضوع أو سلبه عنه في وقت غير معيّن من أوقات وجود الموضوع مقيّداً باللادوام بحسب الذات.
الممكنة الخاصّة وهي التي يحكم فيها بارتفاع الضرورة المطلقة عن جانبي الوجود والعدم جميعاً (الشمسية، 296).	الممكنة الخاصّة ما يحكم فيها بارتفاع الضرورة المطلقة عن جانبي الوجود والعدم جميعاً.
قياس الخلف وهو إثبات المطلوب بإبطال نقيضه (الشمسية، 452).	قياس الخلف إثبات المطلوب بإبطال نقيضه.

(*cont.*)

al-Shamsiyya **and** ***Taḥrīr al-qawāʾid al-manṭiqiyya***	***Maqālīd al-ʿulūm***
التمثيل وهو إثبات حكم في جزئي وجد في جزئي آخر لمعنىً مشترك بينهما (الشمسية، 454).	التمثيل إثبات حكم جزئي في جزئي آخر لمعنىً مشترك بينهما.
أوّليات وهي قضايا تصوّر طرفيها كافٍ في الجزم بالنسبة بينهما [...]	الأوّليات قضايا تصوّر طرفيها كافٍ في الجزم بينهما.
مشاهدات وهي قضايا يحكم بها بقوىً ظاهرة أو باطنة [...]	المشاهدات قضايا يحكم بها بقوىً ظاهرة أو باطنة.
مجرّبات وهي قضايا يحكم بها لمشاهدات متكرّرة مفيدة لليقين [...]	المجرّبات قضايا يحكم بها لمشاهدات متكرّرة مفيدة لليقين.
حدسيات وهي قضايا يحكم بها لحدس قويّ من النفس مفيد للعلم [...]	الحدسيات قضايا يحكم بها لحدس قويّ في النفس مفيد للعلم.
الحدس هو سرعة الانتقال من المبادئ إلى المطالب.	الحدس سرعة الانتقال من المبادئ إلى المطالب.
ومتواترات وهي قضايا يحكم بها لكثرة الشهادات بعد العلم بعدم امتناعها والأمن من التواطؤ عليها (الشمسية، 457).	المتواترات قضايا يحكم بها لكثرة الشهادات بعد العلم بعدم امتناعها والأمن من التواطؤ عليها.
[...] وهو إمّا لمّي وهو الذي يكون الحدّ الأوسط فيه علّة للنسبة في الذهن والعين [...]	البرهان اللمّي ما يكون الحدّ الأوسط فيه علّة للنسبة في الذهن والعين معاً.
وإمّا إنّي وهو الذي يكون الحدّ الأوسط فيه علّة للنسبة في الذهن فقط (الشمسية، 460).	البرهان الإنّي ما يكون الحدّ الأوسط فيه علّة للنسبة في الذهن فقط.

Chapter 12: Philosophy (*al-ḥikma*)

Samarqandī's *al-Ṣaḥāʾif al-ilāhiyya* and Ibn Sīnā's *Risālat al-Ḥudūd* are the main sources for this chapter. More than thirty terms were based on *maqṣad* 1 of *al-Ṣaḥāʾif al-ilāhiyya*, and around twenty definitions are from *Risālat al-Ḥudūd*. Apart from these two works, some of the terms and definitions can be traced in *Risālat Ḥudūd al-ashyāʾ wa-rusūmihā* attributed to al-Kindī, Ibn Sīnā's *ʿUyūn al-ḥikma*, and Khʷārazmī's *Mafātīḥ al-ʿulūm*.

al-Ṣaḥāʾif al-ilāhiyya*/Ibn Sīnā's *Risālat al-Ḥudūd*/ al-Kindī's *Risālat al-Ḥudūd*/ *ʿUyūn al-ḥikma*/ *Mafātīḥ al-ʿulūm	***Maqālīd al-ʿulūm***
الحكمة استكمال النفس الإنسانية بتصوّر الأمور والتصديق بالحقائق النظرية والعملية على قدر الطاقة البشرية (عيون الحكمة، 16).	علم الحكمة [...] وقيل استكمال النفس الإنسانية بالتصوّرات والتصديقات النظرية بقدر الطاقة البشرية.
الفلسفة [...] وحدّوها أيضاً فقالوا الفلسفة معرفة الإنسان نفسه (رسالة الحدود للكندي، 71).	الفلسفة [...] وقيل معرفة الإنسان نفسه
الفلسفة علم الأشياء الأبدية الكلّية [...] (رسالة الحدود للكندي، 71).	الفلسفة [...] وقيل علم الأشياء الأبدية
والبسيط إن كان جزؤه كالكلّ في الاسم والحدّ فهو البسيط العنصري (الصحائف الإلهية، 68).	البسيط العنصري ما كان جزؤه كالكلّ في الاسم والحدّ
إن إدراك الشيء هو أن تكون حقيقته متمثلة عند المدرك يشاهدها ما به يدرك (الصحائف الإلهية، 165).	الإدراك تمثّل حقيقة الشيء عند المدرك بحيث يشاهدها بما به يدرك.
المشاهدة هي أن يدرک الجزئي حاضراً في الخارج (الصحائف الإلهية، 167).	المشاهدة إدراك الجزئي حاضراً في الخارج.
المشاهدة الحقيقية هي إدراك عين الحاضر في الخارج (الصحائف الإلهية، 167).	المشاهدة الحقيقية إدراك عين الحاضر في الخارج.

(*cont.*)

al-Ṣaḥāʾif al-ilāhiyya/Ibn Sīnā's *Risālat al-Ḥudūd*/ al-Kindī's *Risālat al-Ḥudūd*/ *ʿUyūn al-ḥikma*/ *Mafātīḥ al-ʿulūm*	*Maqālīd al-ʿulūm*
[المرتبة] الرابعة عند حصول النظريات لها بحيث لا تعزب ويسمّى حينئذٍ عقلاً مستفاداً. (الصحائف الإلهية، 186).	العقل المستفاد حصول النظريات لها بحيث لا تعزب عنها.
[...] الخلأ عرّفوه بتعريفين: فالأوّل هو أن لا يكون بين الجسمين المتباعدين ما يلاقيانه [...] الثاني: هو أبعاد مجرّدة عن المادّة فإذا حصلت في مادّة حصل الجسم. (الصحائف الإلهية، 212–213).	الخلأ أن لا يكون بين الجسمين المتباعدين ما يلاقيانه. وقيل: هو أبعاد مجرّدة عن المادّة إذا حصلت في مادّة حصل الجسم.
الملأ هو جسم من جهة ما تمانع أبعاده دخول جسم آخر فيه. (رسالة الحدود لابن سينا، 142).	الملأ جسم من جهة ما تمانع أبعاده دخول جسم آخر فيه.
الكمون هو استتار الشيء عن الحسّ (مفاتيح العلوم، 140).	الكمون استتار الشيء عن الحسّ.
[...] كلّ ما بالقوّة فحصوله بالفعل إمّا دفعة أو على التدريج. والأوّل الكون والثاني الحركة. فالحركة هي خروج ما بالقوّة إلى الفعل على التدريج. [...] والأحسن أن يقال: الحركة هي تغيّر أحوال الموجودات (الصحائف الإلهية، 218–219).	الحركة هي خروج ما بالقوّة إلى الفعل على التدريج. وقيل: هي تغيّر أحوال الموجودات. [...] الكون حصول ما بالقوّة إلى الفعل دفعة.
التخلخل ازدياد مقدار الجسم بدون انضمام جسم آخر معه. والتكاثف انتقاص مقدار الجسم بدون انتقاص جسم آخر عنه.	التخلخل ازدياد مقدار الجسم بدون انضمام جسم آخر معه. التكاثف انتقاص مقدار الجسم بدون انتقاص جسم عنه.

(*cont.*)

al-Ṣaḥā'if al-ilāhiyya/Ibn Sīnā's *Risālat al-Ḥudūd*/ al-Kindī's *Risālat al-Ḥudūd*/ *ʿUyūn al-ḥikma*/ *Mafātīḥ al-ʿulūm*	*Maqālīd al-ʿulūm*
والنموّ ازدياد الجسم باتّصال جسم آخر به مداخلاً في أجزائه في الأقطار على نسبة طبيعية [...] والذبول عكسه (الصحائف الإلهية، 221–222).	النموّ ازدياد جسم باتّصال جسم آخر به مداخلاً في أجزائه في الأقطار على نسبة طبيعية. الذبول عكسه.
المكان هو السطح الباطن من الجرم الحاوي المماسّ للسطح الظاهر من الجسم المحويّ. ويقال: مكان للسطح الأسفل الذي يستقرّ عليه جسم ثقيل (رسالة الحدود لابن سينا، 141).	المكان هو السطح الباطن من الجسم الحاوي المماسّ للسطح الظاهر من المحويّ. وقيل: مكان السطح الأسفل الذي يستقرّ عليه شيء ثقيل.
الحركة هي خروج من القوّة إلى الفعل لا في آنٍ واحد (رسالة الحدود لابن سينا، 138).	الحركة هي الخروج من القوّة إلى الفعل لا في آنٍ واحد.
الحركة إمّا سريعة أو بطيئة، والسريعة ما تقطع مسافة أطول في زمان مساوٍ، أو المثل في زمان أقلّ، أو الأطول في الأقلّ. والبطيئة بالعكس (الصحائف الإلهية، 224).	الحركة السريعة ما يقطع مسافة أطول في زمان مساوٍ، أو المثل في زمان أقلّ، أو الأطول في الأقلّ. الحركة البطيئة عكس ذلك.
ثمّ الحركة إمّا مستقيمة أو مستديرة وإمّا مركّبة منهما [...] وهي تسمّى لولبية (الصحائف الإلهية، 224–225).	الحركة الكوكبية حركة مركّبة من المستقيمة والمستديرة.
ثمّ الحركة إمّا بالذات أو بالعرض، لأنّ الحركة إن كانت حاصلة في المتحرّك فهو متحرّك بالذات [كالماشي]، وإن كانت فيما يقارنه فهو متحرّك بالعرض.	المتحرّك بالذات ما يكون الحركة حاصلة فيه. المتحرّك بالعرض ما يكون الحركة فيما يقارنه. الحركة القسرية ما يكون بقوّة حاصلة مستفادة من خارج.

(*cont.*)

al-Ṣaḥāʾif al-ilāhiyya**/Ibn Sīnā's *****Risālat al-Ḥudūd*****/ al-Kindī's *****Risālat al-Ḥudūd*****/ *****ʿUyūn al-ḥikma*****/ *****Mafātīḥ al-ʿulūm***	***Maqālīd al-ʿulūm***
[...] وما بالذات إمّا طبيعية أو قسرية أو إرادية، لأنّ الحركة بالذات لقوّة حاصلة في المتحرّك فتلك القوة إن كانت مستفادة من خارج فهي القسرية [...] وإن كانت من المتحرّك فإن كانت مع القصد فهي إدارية [إرادية؟] وإلّا فالطبيعية (الصحائف الإلهية، 225–226).	الحركة الإرادية ما يكون بقوّة حاصلة من خارج مع القصد. الحركة الطبيعية ما لا يكون بقصد.
حدّ السكون هو عدم الحركة فيما من شأنه أن يتحرّك (رسالة الحدود لابن سينا، 142).	السكون هو عدم الحركة فيما من شأنه أن يتحرّك.
حدّ الآن هو طرف [ظرف؟] موهوم يشترك فيه الماضي والمستقبل من الزمان.	الآن ظرف موهوم يشترك فيه الماضي والمستقبل من الزمان.
وقد يقال: آن لزمان صغير المقدار عند الوهم متّصل بالآن الحقيقي من جنسه (رسالة الحدود لابن سينا، 139).	وقيل: الزمان صغير المقدار عند الوهم متّصل بالآن الحقيقي من جنبيه.
الفلك هو جرم بسيط كريّ غير قابل للكون والفساد، متحرّك بالطبع على الوسط مشتمل عليه (رسالة الحدود لابن سينا، 136).	الفلك جسم بسيط كريّ متحرّك بالطبع على الوسط مشتمل عليه.
حدّ الملك هو جوهر بسيط ذو حياة ونطق عقلي غير مائت وهو واسطة بين البارئ والأجسام الأرضية (رسالة الحدود لإبن سينا، 136).	الملك جوهر بسيط ذو حياة ونطق هو واسطة بين البارئ والأجسام الأرضية.
حدّ الإبداع [...] تأسيس الشيء لا عن شيء ولا بواسطة شيء (رسالة الحدود لابن سينا، 149).	الإبداع إظهار شيء لا من شيء ولا بواسطة شيء.

(*cont.*)

al-Ṣaḥāʾif al-ilāhiyya/Ibn Sīnā's *Risālat al-Ḥudūd*/ al-Kindī's *Risālat al-Ḥudūd*/ *ʿUyūn al-ḥikma*/ *Mafātīḥ al-ʿulūm*	*Maqālīd al-ʿulūm*
حدّ الخلق [...] ويقال خلق لإفادة وجود حاصل عن مادّة وصورة كيف كان (رسالة الحدود لإ بن سينا، 150).	الخلق إفادة وجود حاصل عن مادّة وصورة، كيف كان.
حدّ الاجتماع هو وجود أشياء كثيرة يعمّها معنىً واحد. والافتراق مقابله (رسالة الحدود لابن سينا، 146).	الاجتماع وجود أشياء كثيرة يعمّها معنىً واحد. الافتراق مقابله.
حدّ المتماسّين هما اللذان نهاياتهما معاً في الوضع ليس يجوز أن يقع بينهما شيء ذو وضع (رسالة الحدود لابن سينا، 146).	المتماسّان اللذان نهاياتهما معاً في الوضع، وليس يجوز أن يقع بينهما شيء ذو وضع.
حدّ المتداخل هو الذي يلاقي الآخر بكلّيته حتّى يكفيهما مكان واحد (رسالة الحدود لابن سينا، 146).	التداخل تلاقي الشيء الآخر بكلّيته حتّى يكفيهما مكان واحد.
حدّ المتّصل [...] إنّه ما من شأنه أن يوجد بين أجزائه حدّ مشترك. ورسمه أنّه القابل للانقسام بغير نهاية (رسالة الحدود لابن سينا، 146).	المتّصل ما من شأنه أن يوجد بين أجزائه حدّ مشترك. وقيل: القابل للانقسام بغير نهاية.
حدّ الاتّحاد هو مشترک، فيقال: اتحاد لاشتراك أشياء في محمول واحد ذاتي أو عرضي [...]	الاتّحاد اشتراك أشياء في محمول واحد ذاتي أو عرضي.

(*cont.*)

al-Ṣaḥāʾif al-ilāhiyya*/Ibn Sīnā's *Risālat al-Ḥudūd*/ al-Kindī's *Risālat al-Ḥudūd*/ *ʿUyūn al-ḥikma*/ *Mafātīḥ al-ʿulūm	***Maqālīd al-ʿulūm***
وأحقّ هذا الباب باسم الاتّحاد هو حصول جسم واحد بالعدد من اجتماع أجسام كثيرة لبطلان خاصّياتها لارتفاع حدودها المشتركة وبطلان نهاياتها بالاتّصال (رسالة الحدود لابن سينا، 147).	وقيل: حصول جسم واحد بالعدد من اجتماع أجسام كثيرة لبطلان خاصّيتها لارتفاع حدودها المشتركة وبطلان نهايتها بالاتّصال.
حدّ التتالي هو كون الأشياء التي لها وضع ليس بينها شيء آخر من جنسها (رسالة الحدود لابن سينا، 148).	التتالي كون الأشياء التي لها وضع ليس بينها شيء آخر من جنسها.
حدّ التوالي هو كون شيء بعد شيء بالقياس إلى مبدأ محدود وليس بينهما شيء من بابهما (رسالة الحدود لابن سينا، 148).	التوالي كون شيء بعد شيء بالقياس إلى مبدأ محدود وليس بينهما شيء من بابهما.
حدّ النهاية هي ما به يصير الشيء ذو الكمّية إلى حيث لا يوجد وراءه شيء منه (رسالة الحدود لابن سينا، 139).	النهاية غاية يصير الشيء ذو الكمّية إلى حيث لا يوجد وراءه شيء منه.
حدّ ما لا نهاية له هو كمّ، أيّ أجزاءه أخذ، وجد منه شيئاً خارجاً عنه بعينه غير مكرّر (رسالة الحدود لابن سينا، 139).	الغير المتناهي كمّ، أيّ أجزاءه أخذت، وجد منه شيء خارج عنه بعينه غير مكرّر.

Chapter 13: Astronomy (*al-hayʾa*)

With a few exceptions (for ex. *al-daraja, al-ufuq, ʿarḍ al-balad*), all the terms and definitions in this chapter are taken from Khʷārazmī's *Mafātīḥ ʿulūm*. The table below provides some examples.

Mafātīḥ al-ʿulūm	*Maqālīd al-ʿulūm*
علم الهيئة هو معرفة تركيب الأفلاك وهيئتها وهيئة الأرض (215).	علم الهيئة معرفة تراكيب الأفلاك والأرض.
الفلك المستقيم هو معدّل النهار وهو الدائرة العظمى التي تحيط على قطبي السماء اللذين عليهما يتحرّك من المشرق إلى المغرب دورة في كلّ يوم وليلة (215).	الفلك المستقيم دائرة عظمى تحيط بقطبي السماء اللذين عليهما يتحرّك من المشرق إلى المغرب في كلّ يوم وليلة دورة.
خطّ الاستواء من الأرض هو الخطّ الذي يقابل معدّل النهار (215).	خطّ الاستواء ما يقابل معدّل النهار.
فلك البروج هو الدائرة التي ترسمها الشمس بسيرها من المغرب إلى المشرق في سنة واحدة (215).	فلك البروج دائرة ترسمها الشمس بسيرها من المشرق إلى المغرب في السنة.
دائرة الارتفاع هي التي تمرّ بقطبي الأفق (216).	دائرة الارتفاع ما تمرّ بقطبي الأفق.
قوس الارتفاع قطعة من تلك الدائرة (216).	قوس الارتفاع قطعة من دائرة الارتفاع.
الميل هو بعد الشمس أو الكواكب من معدّل النهار (216).	الميل بعد الشمس أو الكواكب عن معدّل النهار.
سعة المشرق للشمس هو من الأفق ما بين معدّل النهار وبين مطلعها (216).	سعة المشرق للشمس من الأفق ما بين معدّل النهار وبين مطلعها.
مطالع الفلك المستقيم هي ما يطلع مع قسيّ فلك البروج من معدّل النهار في خط الاستواء (219).	مطالع الفلك المستقيم ما يطلع مع قسيّ فلك البروج من معدّل النهار في خطّ الاستواء.
مطالع البلد من البلدان هي ما يطلع مع قسيّ فلك البروج من أفق ذلك البلد (219).	مطالع البلد ما يطلع مع قسيّ فلك البروج من أفق ذلك البلد.
قوس النهار هي القوس التي فوق الأرض من الدائرة الموازية لمعدّل النهار التي فيها تدور الشمس في يوم واحد من الأيام (219–220).	قوس النهار هي القوس فوق الأرض من الدائرة الموازية لمعدّل النهار التي فيها تدور الشمس.

(*cont.*)

Mafātīḥ al-ʿulūm	*Maqālīd al-ʿulūm*
قوس اللّيل ما يبقى لتمام تلك الدائرة (220).	قوس الليل ما يبقى من تمام تلك الدائرة.
الجوزهر هو النقطتان اللّتان تتقاطع عليهما الدائرتان من الأفلاك تسمّيان العقدتين (220).	الجوزهر النقطتان اللتان يتقاطع عليهما الدائرتان من الأفلاك.
الأوج هو أرفع موضع من الفلك الخارج المركز (221).	الأوج أرفع موضع في الفلك الخارج المركز.
منطقة البروج هي نطاق البروج ووسط البروج الذي فيه مسير الشمس (221).	منطقة البروج وسط البروج الذي فيه مسير الشمس.
سير الطول للكوكب هو سيره في نطاق البروج (221).	سير الطول سير الكواكب في نطاق البروج إلى ما يلي قطب الشمال أو الجنوب.
فلك الأوج هو الخارج المركز (222).	فلك الأوج هو الخارج عن المركز.
المركز يعني به سير مركز فلك التدوير في الفلك الخارج المركز (223).	المركز سير مركز فلك التدوير في الفلك الخارج المركز.
القبّة وسط الأرض (218).	القبّة عبارة عن وسط الأرض.
البركسيس هو اختلاف المنظر (222).	البراكيس اختلاف المناظر.
المعمورة من الأرض ربعها الذي على مهبّ الشمال (217).	المعمورة ربع الأرض الذي يلي مهبّ الشمال.

Chapter 14: Geometry (*al-handasa*)

Most of the terms and definitions in this chapter are taken from Khʷārazmī's *Mafātīḥ ʿulūm*. The table below provides some examples. However, the author must have consulted one other source at least, since a large number of the terms and definitions are not from the *Mafātīḥ ʿulūm*.

Mafātīḥ al-ʿulūm	*Maqālīd al-ʿulūm*
المصادرة ما يصدّر به الكتاب أو الباب من أبواب الهندسة من مقدّمات المسألة (203).	المصادرة ما يصدّر الكتاب أو الباب به من أبواب الهندسة.
الشكل القطّاع بفتح القاف وتشديد الطاء قطعة من دائرة رأسها إمّا على مركزها وإمّا على محيطها مثل هذين الشكلين (207).	الشكل القطّاع قطعة دائرة رأسها إمّا على مركزها وإمّا على محيطها مثل هذين الشكلين.
البسيط المقبّب الكريّ ما كان على شكل الكرة (207).	البسيط المقبّب الكري ما كان على شكل الكرة.
البسيط الأسطوانيّ ما كان على شكل الأسطوانة يبتدئ من دائرة وينتهي إلى دائرة (207).	البسيط الأسطواني ما يبتدئ من دائرة و ينتهي إلى دائرة.
الشكل الناريّ هو جسم يحيط به أربعة سطوح مثلّثات متساوية الأضلاع (207).	الشكل الناري جسم تحيط به أربعة سطوح مثلّثات متساوية الأضلاع.
الشكل الهوايي هو جسم يحيط به ثمانية سطوح مثلّثات متساوية الأضلاع والزوايا (208).	الشكل الهوايي جسم تحيط به ثمانية سطوح مثلّثات متساوية الأضلاع والزوايا.
الشكل المائيّ هو جسم يحيط به عشرون مثلّثاً متساوية الأضلاع والزوايا (208).	الشكل المائي جسم يحيط به عشرون مثلّثاً متساوية الأضلاع والزوايا.
الشكل الفلكيّ هو جسم يحيط به اثنى عشر سطحاً مخمّسات متساوية الأضلاع والزوايا (208).	الشكل الفلكي جسم يحيط به اثنى عشر سطحاً مجمّساً متساوية الأضلاع والزوايا.
الشكل اللبنيّ جسم مربّع يكون بعدان من أبعاده متساويين والثالث أصغر على شكل اللّبنة المربّعة (208).	الشكل اللبني جسم مربّع يكون بُعدان من أبعاده متساويين والثالث أصغر من شكل المربّعة.
الشكل العموديّ جسم مربّع يكون بعدان من أبعاده متساويين والثالث أعظم (208).	الشكل العمودي مربّع يكون بُعدان من أبعاده متساويين والثالث أعظم.

(*cont.*)

Mafātīḥ al-ʿulūm	*Maqālīd al-ʿulūm*
الكرة شكل مجسّم يحيط به بسيط واحد في داخله نقطة كلّ الخطوط المستقيمة الخارجة من تلك النقطة إلى بسيطها متساوية وتلك النقطة مركزها (208).	الكرة [...] وقيل: جسم يحيط به بسيط واحد في داخله نقطة، كلّ الخطوط [المستقيمة] الخارجة من تلك النقطة التي بسيطها متساوية، وتلك النقطة مركزها، وينتهي إلى بسيطها.
محور الكرة قطرها الذي تتحرّك عليه الكرة وهو ثابت (208–209).	محور الكرة قطر نصف الدائرة المثبت. وقيل: قطرها الذي يتحرّك عليه وهو ثابت.
قطبا الكرة طرفا المحور (209).	قطب الكرة طرفا نصف الدائرة المثبت.
البيضة شكل مجسّم يحيط به بسيط واحد وتحدث عن قطعة أقلّ من نصف دائرة إذا صيّر طرفاها كالمحور وأديرت إلى أن ترجع إلى حيث ابتدأت منه (209).	البيضية شكل مجسّم يحيط به بسيط واحد ويحدث عن قطعة أقلّ من نصف دائرة إذا صيّر طرفها كالمحور، وأدير إلى أن يرجع إلى حيث ابتدأ منه.
الحلقة هي جسم يحيط به بسيط واحد مستدير في داخله مكان يمكن أن تقع فيه كرة (209).	الحلقة جسم يحيط به بسيط واحد مستدير في داخله مكان يمكن أن يقع فيه كرة.
الأسطوانة جسم يبتدئ من دائرة وينتهي إلى دائرة متساوية لها يحيط بها بسيط أسطواني (209).	الأسطوانة جسم يبتدئ من دائرة وينتهى إلى دائرة مساوية لما يحيط به بسيط أسطوانيّ.

Chapter 15: Arithmetic (*al-ḥisāb*)

Most of the terms and definitions in this chapter are taken from Khʷārazmī's *Mafātīḥ ʿulūm*. The table below provides some examples. However, the author must have consulted at least one other source, since a large number of the terms and definitions are not from the *Mafātīḥ ʿulūm*.

Mafātīḥ al-ʿulūm	*Maqālīd al-ʿulūm*
العدد الدوائريّ ما كان بدؤه ونهايته شيئاً واحداً (190–191).	العدد الدوائريّ ما كان بدؤه ونهايته شيئاً واحداً.
المال كلّ ما يجتمع من ضرب عدد في نفسه. (198).	المال كلّ ما يجتمع من ضرب عدد في نفسه.
الجذر كلّ ما تضربه في نفسه (198).	الجذر كلّ ما يضرب في نفسه.
الجذر المطلق هو المنطوق به وهو ما يعرف حقيقة مقداره ويمكن أن ينطق به (198).	الجذر المطلق ما يعرف حقيقة مقداره ويمكن أن ينطق به.
الجذر الأصمّ الذي لا سبيل إلى علم حقيقته بالعدد (198).	الجذر الأصمّ ما لا سبيل إلى علم حقيقته بالعدد.
منها مثلّثة القواعد [...] وتتولّد من تراكم المثلّثات (189).	مثلّثة القواعد ما يتولّد من تراكم المثلّثات.
منها مربّعة القواعد [...] فتتولّد من تراكم المربّعات (189).	مربّعة القواعد ما يتولّد من تراكم المربّعات.
ذو الاسمين ما لا يمكن أن ينطق به بلفظ واحد (199).	ذو الاسمين ما لا يمكن أن ينطق بلفظ واحد.
المكعّب هو المال إذا ضرب في ضلعه أي جذره فالمبلغ هو المكعّب وذلك الجذر هو الكعب (199).	المكعّب هو المال إذا ضرب في جذره، فالمبلغ هو المكعّب، وذلك الجذر هو الكعب.
مال المال هو المال إذا ضرب في نفسه فإنّ المجتمع هو مال المال وكذلك إذا ضرب المكعّب في كعبه صار مال المال (199).	مال المال ما إذا اجتمع من ضرب المال في نفسه، وكذلك إذا ضرب المكعّب في كعبه.
الشيء في كلام أهل الجبر والمقابلة هو الجذر المجهول (199–200).	الشيء في كلام أهل الجبر هو الجذر المجهول.

(*cont.*)

Mafātīḥ al-ʿulūm	*Maqālīd al-ʿulūm*
القسمة أخذ حصّة الواحد من المقسوم عليهم من المقسوم (198).	القسمة أخذ حصّة الواحد من المقسوم عليه.

Chapter 16: Finance (*al-istīfāʾ*)

Almost all the terms and definitions in this chapter are taken from Khʷārazmī's *Mafātīḥ ʿulūm* (Chapter Four on secretary's arts). The table below provides some examples.

Mafātīḥ al-ʿulūm	*Maqālīd al-ʿulūm*
قانون الخراج أصله الذي يرجع إليه وتبنى الجباية عليه (54).	قانون الخراج أصله الذي يرجع إليه ويبنى الجباية عليه.
الأوارج إعراب أواره ومعناه بالفارسية المنقول؛ لأنّه ينقل إليه من القانون ما على إنسان إنسان ويثبت فيه ما يؤدّيه دفعة بعد أخرى إلى أن يستوفى ما عليه (54).	الأوارج ما ينقل من القانون على إنسان إنسان ويثبت ما يؤدّيه دفعة بعد أخرى إلى أن يستوفى ما عليه.
الروزنامج تفسيره كتّاب اليوم؛ لأنّه يكتب فيه ما يجري كلّ يوم من استخراج أو نفقة أو غير ذلك (54).	الروزنامج كتّاب اليوم يثبت فيه ما يجري من استخراج أو نفقة.
الختمة كتّاب يرفعه الجهبذ في كلّ شهر بالاستخراج والجمل والنفقات والحاصل كأنّه يختم الشهر به (54).	الختمة حساب يرفعه الجهبذ في كلّ شهر بالاستخراج والجمل والنفقات.
الختمة الجامعة تعمل كلّ سنة كذلك (54).	الختمة الجامعة ما يعمل هكذا لكلّ سنة.
التأريج قيل لفظة فارسية ومعناه النظام؛ لأنه كسواد يعمل للعقد لعدّة أبواب يحتاج إلى علم جملها (54).	التأريج سواد يعمل [للعقد] لعدّة أبواب يحتاج إلى علم جملها.

(*cont.*)

Mafātīḥ al-ʿulūm	*Maqālīd al-ʿulūm*
العريضة شبيهة بالتأريج إلّا أنّها تعمل لأبواب يحتاج إلى أن يعلم فضل ما بينها فينقص الأقلّ من الأكثر من بابين منها، ويوضع ما يفضل في باب ثالث وهو الباب المقصود الذي تعمل العريضة لأجله (55).	العريضة شبيهة بالتأريج إلا أنّها لأبواب يحتاج أن يعلم فصل ما بينهما فينقص الأوّل من الأكثر من بابين منها، ويوضع ما يفصل في باب واحد، والمقصود من العريضة هو.
البراءة حجة يبذلها الجهبذ أو الخازن للمؤدّي بما يؤدّيه إليه (55–56).	البراءة حجّة يبذلها الجهبذ أو الخازن للمؤدّي ما يؤدّى إليه.
الموافقة والجماعة حساب جامع يرفعه العامل عند فراغه من العمل ولا يسمّى موافقة ما لم يرفع باتّفاق بين الرافع والمرفوع إليه، فإن انفرد به أحدهما دون أن يوافق الآخر على تفصيلاته سمّي محاسبة (56).	الموافقة حساب جامع يرفعه العامل عند فراغه من العمل باتّفاق بين الرافع والمرفوع [إليه]. الجماعة هي الموافقة. المحاسبة ما تفرّد بها الرافع والمرفوع [إليه] على تفصيلاته.
من دفاتر ديوان الجيش الجريدة السوداء وهي تكسر لقيادة قيادة في كلّ سنة بأسامي الرجال وأنسابهم وأجناسهم وحلاهم ومبالغ أرزاقهم وقبوضهم وسائر أحوالهم، وهو الأصل الذي يرجع إليه في هذا الديوان في كلّ شيء (56).	الجريدة ما يكتب في كلّ سنة بأسامي الرجال وكناهم وأنسابهم وحلاهم ومبالغ أرزاقهم وقبوضهم.
الرجعة حساب يرفعه المعطي في بعض العساكر بالنواحي لطمع واحد إذا رجع إلى الديوان (56).	الرجعة حساب يرفعه المعطى في بعض العساكر.
الرجعة الجامعة يرفعها صاحب ديوان الجيش لكلّ طمع من صنوف الإنفاق (56).	الرجعة الجامعة ما يرفعها صاحب ديوان الجيش لكلّ جمع من صنوف الأرزاق.

(*cont.*)

Mafātīḥ al-ʿulūm	*Maqālīd al-ʿulūm*
الصكّ عمل يعمل لكل طمع يجمع فيه أسامي المستحقين وعدّتهم ومبلغ ما لهم ويوقع السلطان في آخره باطلاق الرزق لهم (56).	الصكّ ما يعمل لكلّ طمع يجمع فيه أسامي المستحقّين وعددهم ومبلغ مالهم ليوقّع السلطان عليه بإطلاق الرزق عليهم.
المؤامرة عمل تجمع فيه الأوامر الخارجة في مدة أيام الطمع ويوقع السلطان في آخره بإجازة ذلك، وقد تعمل المؤامرة في كلّ ديوان تجمع جميع ما يحتاج إليه من استئمار واستدعاء توقيع (56–57).	المؤامرة ما يجمع فيه الأوامر الخارجة في مدّة أيّام الطمع ليوقّع السلطان فيه إلى استئمار واستدعاء توقيع.
الاستقرار عمل يعمل لما يستقرّ عليه من الطمع بعد الإثبات والفكّ والوضع والزيادة والحطّ والنقل والتحويل ونحو ذلك (57).	الاستقرار ما يعمل لما استقرّ عليه أمر الطمع بعد الإثبات والفكّ والوضع والزيادة والحطّ والتحويل والنقل وغيرها.
أمّا السجّل فكتاب يكتب للرسول أو المخبّر أو الرحّال أو غيرهم بإطلاق نفقته حيث بلغ فيقيمها له كلّ عامل يجتاز به، والسجّل أيضاً المحضر يعقده القاضي بفصل القضاء (57).	السجّل ما يكتب لرسول أو لمجمّز أو رحّال بإطلاق نفقته حيث بلغ إلى كلّ عامل يجتاز به، ويقال أيضاً لما يكتبه القاضي لتفصيل القضاء.

Chapter 17: Music (*al-mūsīqā*)

Most of the terms and definitions in this chapter are taken from Khʷārazmī's *Mafātīḥ ʿulūm*. The table below provides some examples. However, the author must have consulted at least one other source, since a number of the definitions (for ex. *al-zāʾid, al-īqāʿāt, khafīf al-thaqīl, al-raml*) are not from the *Mafātīḥ ʿulūm*.

Mafātīḥ al-ʿulūm	*Maqālīd al-ʿulūm*
النغمة صوت غير متغيّر إلى حدّة ولا ثقل (240).	النغمة صوت غير متغيّر إلى حدّة.
البعد صوت يبتدأ فيه بنغمة، ويثنّى فيه بنغمة أخرى (241).	البعد صوت بين النغمتين يبتدأ فيه بنغمة، ويثنّى فيه بنغمة أخرى.
مراتب حدّة الصوت أو ثقله تسمّى الطبقات (241).	الطبقات مراتب حدّة الصوت أو ثقله.
البعد ذو الخمس [...] هو مثل ما بين مطلق البم إلى سبّابة المثلث [...] (241).	البعد ذو الخمس هو مثل ما بين مطلق البم إلى سبّابة المثلث.
البعد ذو الأربع [...] هو ما بين مطلق البم إلى خنصره، وهو ربع الوتر [...] (242).	البعد ذو الأربع هو ما بين مطلق البم إلى خنصره، وهو ربع الوتر.
البعد الطنيني والمدّة والعودة هو ما بين المطلق والسبّابة [...] (242).	البعد الطنيني هو ما بين المطلق والسبّابة. المدّة والعودة كذلك.
الفضلة والبقيّة هي بعد ما بين البنصر والخنصر، أو ما بين السبّابة والوسطى أو ما بين السبّابة ووسطى الفرس. وهو نصف المدّة بالتقريب (242–243).	الفضلة بعد ما بين البنصر والخنصر، أو ما بين السبّابة ووسطى الفرس. وقيل هي نصف المدّة بالتقريب والبقية مثلها.
الإرخاء هو نصف الفضلة بالتقريب (243).	الإرخاء نصف الفضلة.
الدساتين هي الرباطات التي توضع الأصابع عليها واحدها دستان. والدستان أيضاً اسم لكلّ لحن من الألحان المنسوبة إلى باربد (238).	الدستان اسم لكلّ من الألحان المنسوبة إلى فاربذ. وقيل: [...] وقيل: هي الرباطات التي يوضع عليها الأصابع.
مشط العود هو الشبيه بالمسطرة التي يشدّ عليها الأوتار من تحت أنف العود [...] (239).	مشط العود مسطرة تشدّ عليها الأوتار.
عينا العود هما النقبتان اللتان على وجهه (239).	عينا العود الثقبتان اللتان على وجهه.
الإ بريق اسم لعنق العود بما فيه من الآلات (239).	الإ بريق عنق العود بما فيه من الآلات.
المضراب هو الذي يضرب به الأوتار (239).	المضراب ما يضرب به الأوتار.

(*cont.*)

Mafātīḥ al-ʿulūm	*Maqālīd al-ʿulūm*
الجسّ هو نقر الأوتار بالسبّابة والإبهام دون المضراب يشبه ذلك بجسّ (239–240).	الجسّ نقر الأوتار بالسبّابة والإبهام دون المضراب. الحزق مدّ الوتر.
الحزق هو مدّ الوتر. ونقيضه الإرخاء (240).	الحزق مدّ الوتر. الإرخاء ضدّه.
أوتار العود الأربعة، أغلظها البمّ؛ والذي يليه المثلث [...] والذي يلي المثلث المثنى [...] والرابع هو الزير وهو أدقّها (238).	البمّ أغلظ الأوتار الأربعة من العود. المثلث الذي يليه. المثنيّ الذي يلي المثلث. الزير أدقّ الأربعة.
الأجناس ثلاثة: الطنينيّ ويسمّى القويّ والمقويّ وهو أن يقسم البعد ذو الأربع بمدّة ومدّة ونصف مدّة مثل نغمة المطلق ثمّ السبّابة ثمّ البنصر ثمّ الخنصر. الجنس الثاني اللويّ والملوّن وهو أن يقسّم البعد ذو الأربع بنصف مدّة ونصف مدّة وثلث مدّة وثلاثة أنصاف مدّة. والجنس الثالث ويسمّى التأليفي والناظم والراسم وهو أن يقسّم البعد ذو الأربع بربع مدّة وربع مدّة ومدّتين (243).	الطنيني أن يقسّم البعد ذو الأربع بمدّة ونصف مدّة، ثمّ نغمة المطلق، ثمّ السبّابة، ثمّ البنصر، ثمّ الخنصر. القويّ والمقويّ والرجلي كذلك. اللوني أن يقسّم البعد ذو الأربع بنصف مدّة، ونصف مدّة وثلاثة أنصاف مدّة. الملوّن والخنثويّ كذلك. التأليفي أن يقسّم البعد ذو الأربع بربع مدّة ومدّتين. النسوي والناظم والراسم كذلك.

Chapter 18: Astrology (*al-nujūm*)

Most of the terms and definitions in this chapter are taken from Khʷārazmī's *Mafātīḥ ʿulūm*. The table below provides some examples. However, the author must have consulted at least one other source, since the definitions of a number of the terms are not identical to those in the *Mafātīḥ ʿulūm*.

Mafātīḥ al-ʿulūm	*Maqālīd al-ʿulūm*
الزيج كتّاب منه يحسب سير الكواكب ومنه يستخرج التقويم (219).	الزيج كتّاب يحسب منه سير الكواكب ليستخرج منه التقويم.
الزائجة هي صورة مربّعة أو مدوّرة تعمل لمواضع الكواكب في الفلك لينظر فيها عند الحكم لمولد أو غيره (219).	الزايجة صورة مربّعة أو مدوّرة تعمل لمواقع الكواكب في الفلك لينظر فيها عند الحكم لمولد الطفل أو غيره.
الهيلاج [...] وهي أدلّة العمر (230)	الهيلاج عبارة عن دليل العمر.
بيت الكوكب برج ينسب إليه (225).	بيت الكوكب بيت ينسب إليه.
شرف الكوكب درجة في برج ينسب اليه (225). [...] الهبوط مقابل الشرف (227).	شرف الكوكب درجة في برج ينسب اليه. الهبوط مقابله.
الوبال هو البرج المقابل للبيت (226–227).	الوبال هو البرج المقابل لبيت الكوكب.
المثلّثة كلّ ثلاثة أبرج تكون على طبيعة واحدة تنسب إلى ثلاثة كواكب [...] (225).	المثلّثة كلّ ثلاثة أبرج تكون على طبيعة واحدة ينسب إليه ثلاثة كواكب.
الحيّزان يكون الكوكب الذكر في برج ذكر بالنهار فوق الأرض وبالليل تحت الأرض، أو يكون الكوكب الأنثى في برج أنثى بالنهار تحت الأرض وبالليل فوق الأرض فيقال هو في حيّزه (229).	حيّز الكوكب كونه إذا كان ذكرياً نهارياً فوق الأرض بالنهار، وإذا كان ليلياً مؤنّثاً تحت الأرض بالليل.

(*cont.*)

Mafātīḥ al-ʿulūm	*Maqālīd al-ʿulūm*
الآبار درج في البروج إذا بلغتها الكواكب نحست فيها. (227).	الآبار درج في البروج إذا بلغتها الكواكب نحست فيها.
الابتزاز أن يكون للكوكب حظوظ كثيرة في البرج فيقال هو مبتزّ عليه. (229).	الابتزاز أن يكون للكوكب حظوظ كثيرة في البرج.
الاستعلاء أن يكون الكوكب في البرج العاشر من الآخر (229).	الاستعلاء أن يكون الكوكب في البرج العاشر من العاشر الآخر.
الحصار أن يكون الكوكب مضغوطاً بين نحسين، أحدهما أمامه والآخر وراءه (229).	الحصار أن يكون مضغوطاً بين نحسين.
الفردار قسمة العمر بين الكواكب السبعة لكلّ كوكب منها سنون معلومة يقال لها سنو الفردار (231).	الفردار قسمة للقمر بين الكواكب السبعة.
معنى التسيير أن ينظر كم بين الهيلاج وكم بين السعد أو النحس فيؤخذ لكلّ درجة سنة فيقال تصيبه السعادة أو النكبة إلى كذا وكذا سنة (230).	التسيير أن ينظر حكم بين الهيلاج وبين السعد والنحس فيؤخذ لكلّ درجة سنة.
معنى التربيع أن يصير منه على ربع الفلك (232).	التربيع أن يصير الكوكب من آخر على ربع الفلك.
التثليث أن يصير منه على ثلث الفلك (232).	التثليث أن يصير منه على ثلث الفلك.
التسديس أن يصير منه على سدس الفلك (232).	التسديس أن يصير منه على سدس الفلك.
المقابلة أن يصير منه على نصف الفلك (232).	المقابلة أن يصير منه على نصف الفلك.
الاجتماع يعنى به المحاقّ (232).	الاجتماع يعبّر به عن المحاق.
القِران يعنى به اجتماع زحل والمشتري خاصّة (232).	القِران يعنى به اجتماع زحل والمشتري خاصّة.

Chapter 19: Medicine (*al-ṭibb*)

One of the main sources for this chapter is Ibn Nafīs' *al-Ṣaydaliyya al-mujarraba*. Altogether, thirty-nine terms and definitions are taken from this work. The author also consulted Ibn Sīnā's *Qānūn* (Book 2. 1), and Khʷārazmī's *Mafātīḥ al-ʿulūm*. The table below provides examples of the terms adopted from the above-mentioned works. However, the author must have had at least one other source at his disposal, which has remained unknown to us.

al-Ṣaydaliyya al-mujarraba/ al-Qānūn/Mafātīḥ al-ʿulūm	*Maqālīd al-ʿulūm*
إنّ الطبّ علم يتعرّف منه أحوال بدن الإنسان من جهة ما يصحّ ويزول عن صحّته ليحفظ الصحّة حاصلة ويستردّها زائلة (القانون، ج1، 13).	علم الطبّ ما يعرف منه أحوال بدن الإنسان من جهة ما يصحّ ويزول عن صحّته لتحفظ الصحّة حاصلة وتستردّ زائلة.
[...] ما يكون التعليم فيه مقيّد الاعتقاد فقط، من غير أن يتعرّض لبيان كيفية عمل (القانون، ج 1، 13).	الطبّ النظري ما لا يتعرّض فيه لبيان كيفية العمل.
الصحّة ملكة أو حالة تصدر عنه الأفعال من الموضوع لها سليمة (القانون، ج 1، 13).	الصحّة هيئة بدنية يكون الأفعال بها لذاتها سليمة.
الرياضة هي حركة إرادية تضطرّ إلى التنفّس العظيم المتواتر (القانون، ج 1، 221).	الرياضة حركة إرادية تضطرّ إلى التنفّس العظيم المتواتر.
التدبير التصرّف في الأسباب الضرورية المعدودة التي هي جارية في العادة. (القانون، ج 1، 267).	التدبير التصرّف في الأسباب الضرورية.
من الدلك دلك الاستعداد وهو قبل الرياضة [...] ومنه دلک الاسترداد وهو بعد الرياضة. (القانون، ج 1، 225).	الاستعداد تقديم الدلك على الرياضة. الاسترداد تأخيره عنها.

(*cont.*)

Maqālīd al-ʿulūm	*al-Ṣaydaliyya al-mujarraba/ al-Qānūn/Mafātīḥ al-ʿulūm*
النبض حركة وضعية للشرايين قبضاً وبسطاً لتعديل الروح بالنسيم وإخراج فضلاته.	هي حركة وضعية للشرايين قبضاً وبسطاً لتعديل الروح بالنسيم وإخراج فضلاته. (الصيدلية المجرّبة، 59).
القوّة الطبيعية ما يتصرّف في الغذاء لبقاء الشخص، وبعد تصرّفها لبقاء النوع.	أمّا القوى الطبيعية، [...] جنس يتصرّف في الغذاء لبقاء الشخص [...] وجنس يتصرّف في الغذاء لبقاء النوع. (القانون، ج 1، 95).
الغاذية ما تحيل الغذاء إلى مشابهة المغتذي.	القوّة الغاذية فهي التي تحيل الغذاء إلى مشابهة المغتذي. (القانون، ج 1، 95).
النامية ما تزيد في أقطار البدن على نسبة مخصوصة حتّى تبلغ به كمالها المقداري ثمّ يبطل فعلها.	أمّا النامية فهي الزائدة في أقطار الجسم على التناسب الطبيعي ليبلغ تمام النشء بما يدخل فيه من الغذاء. (القانون، ج 1، 95).
الملطّف ما يجعل المادّة أرقّ.	الملطّف ما يجعل المادّة أرقّ (الصيدلية المجرّبة، 94).
المحلّل ما يهيّئ المادّة للتبخير.	المحلّل ما يهيّئ المادّة للتبخير (الصيدلية المجرّبة، 95).
المنشّف ما إذا لاقته مائية غاصت في مسامّه، فلا يظهر فيه أثر.	المنشّف ما إذا لاقته مائية غاضت في مسامّه، فلا يظهر فيه أثر (الصيدلية المجرّبة، 94).
الجالي ما يجرّد الرطوبة اللزجة عن مسامّ العضو.	الجالي ما يجرّد الرطوبة اللزجة عن مسامّ العضو (الصيدلية المجرّبة، 95).
المخشّن ما يجعل أجزاء سطح العضو مختلفة الوضع بعد ملاسة طبيعية أو عارضة عن مادّة لزجة.	المخشّن ما يجعل أجزاء سطح العضو مختلفة الوضع بعد ملاسة طبيعية أو عارضة عن مادّة لزجة (الصيدلية المجرّبة، 95).
المفتّح ما يخرج المادّة السادّة عن المجرى إلى الخارج.	المفتّح ما يخرج المادّة السادّة عن المجرى إلى الخارج (الصيدلية المجرّبة، 95).

(*cont.*)

al-Ṣaydaliyya al-mujarraba/ al-Qānūn/Mafātīḥ al-ʿulūm	*Maqālīd al-ʿulūm*
المرخّي ما يليّن العضو بحرارته ورطوبته (الصيدلية المجرّبة، 95).	المرخّي ما يليّن العضو بحرارته ورطوبته.
المنضّج ما يعدل قوام الخلط ويهيّئه للدفع (الصيدلية المجرّبة، 95).	المنضِّج ما يعدل قوام الخلط ويهيّئه للدفع.
الهاضم ما يفيد الغذاء سرعة إنضاج (الصيدلية المجرّبة، 95).	الهاضم ما يفيد الغذاء سرعة إنضاج.
المقطّع ما يقسّم المادّة إلى أجزاء صغار، وإن بقيت على غلظها (الصيدلية المجرّبة، 95).	المقطّع ما يقسّم المادّة إلى أجزاء صغار، وإن بقيت على غلظها.
الجاذب ما يحرّك المادّة إلى موضعه (الصيدلية المجرّبة، 95).	الجاذب ما يحرّك المادّة إلى موضعه.
اللاذع ما يفرّق بقوّة نفّاذة اتّصال العضو في مواضع لا تحسّ بانفرادها بل جملتها (الصيدلية المجرّبة، 95).	اللاذع ما يفرّق اتّصال العضو بقوّة نفّاذة في مواضع لا تحسّ بانفرادها بل بجملتها.
الباسليق وهو في اليد عند المرفق في الجانب الإنسيّ إلى ما يلي الإبط (مفاتيح العلوم، ص 153)	الباسليق عرق في اليد عند المرفق في الجانب الإنسي الأيسر إلى ما يلي الإبط.
القيفال عند المرفق أيضا في الجانب الوحشيّ (مفاتيح العلوم، 153).	القيفال هو عند المرفق في الجانب الوحشي.
الأكحل بين الباسليق والقيفال (مفاتيح العلوم، 153).	الأكحل بين الباسليق والقيفال.
الودجان عرقان في العنق أحدهما الودج الظاهر والآخر الودج الغائر (مفاتيح العلوم، 153).	الودجان العرقان في العنق من جانبي الحلقوم.
حبل الذراع عرق في ظاهر الساعد وهو من شعب القيفال (مفاتيح العلوم، 153–154).	حبل الذراع في ظاهر الساعد من شعب القيفال.

(*cont.*)

al-Ṣaydaliyya al-mujarraba/ al-Qānūn/Mafātīḥ al-ʿulūm	*Maqālīd al-ʿulūm*
الأسيلم عرق بين الخنصر والبنصر وهو من شعب الباسليق وهو معرّب (مفاتيح العلوم، 154).	الأسيلم عرق بين الخنصر والبنصر.
الصّافن عرق في الساق يظهر عند الكعب الداخل في الجانب الإنسيّ (مفاتيح العلوم، 154).	الصافن عرق في الساقين يظهر عند الكعب الداخل في الجانب الإنسيّ.
عرق النّسا بفتح النون مقصور قبالة الصافن في الجانب الوحشيّ (مفاتيح العلوم، 154).	عرق النسا قبالة الصافن من الجانب الوحشي.
البوّاب معىً متّصل بالمعدة من أسفل ينضمّ عند دخول الطعام المعدة إلى أن ينهضم فحينئذ ينفتح بإذن الله تعالى ولذلك سمّي البوّاب (مفاتيح العلوم، 155).	البوّاب معىً متّصلة بالمعدة من أسفل ينضمّ عند دخول الطعام إلى المعدة إلى أن ينهضم ثمّ ينفتح.

Chapter 20: Ethics (*al-akhlāq*)

All the terms in this chapter are taken from the first five chapters (*fuṣūl*) of Rāghib Iṣfahānī's *al-Dharīʿa ilā makārim al-sharīʿa*. The table below provides some examples.

al-Dharīʿa ilā makārim al-sharīʿa	*Maqālīd al-ʿulūm*
قوّة النزوع وبها يكون الطلب للموافق والهرب من المخالف والرضا والغضب والإيثار والكراهة (74).	قوّة النزوع ما بها يكون الطلب للموافق والهرب من المخالف والرضا والغضب والإيثار والكرامة.
قوّة التفكّر وبها يكون النطق والعقل والحكمة والرويّة والتدبير والمهنة والرأي والمشورة (74).	قوّة التفكّر ما بها يكون النطق والعقل والعلم والحكمة والرويّة والمهنة والرأي والمشورة.

(*cont.*)

al-Dharīʿa ilā makārim al-sharīʿa	*Maqālīd al-ʿulūm*
قوّة الغذاء وبها النشأة والتربية والولادة (73).	قوّة الغذاء ما بها النشؤ والتربية والولادة.
قوّة الحسّ وبها الإحساس واللذّة والألم (74).	قوّة الحسّ ما بها يكون الإحساس واللذّة والألم.
والخَلق في الهيئات والأشكال والصور المدرَكة بالبصر (96).	الخلق اسم جامع للهيئات والأشكال والصور المدرَكة بالبصر.
والضريبة اعتبارا بضرب الدراهم (96).	الضريبة هكذا، اعتباراً بضرب الدرهم.
والغريزة اعتباراً بما غرز عليه (96).	الغريزة لما غرز عليه.
والشيمة اسم للحالة التي عليها الغريزة اعتباراً بالشامّة اللتي في أصل الخلقة (96).	الشيمة الحالة التي عليها الغريزة اعتباراً بالشامّة اللتي في أصل الخلقة.
السجية اسم لما يسجى عليه الانسان (96).	السجية ما سجي عليه الانسان.
أمّا العادّة فاسم لتكرير الفعل أو الانفعال (97).	العادّة تكرير الفعل والانفعال.
النهي [...] جعل اسماً للعقل الذي انتهى من المحسوسات إلى معرفة ما فيه من المعقولات (137).	النهي العقل المنتهي من المحسوسات إلى معرفة ما فيها من المعقولات.
الحجر [...] وهو اسم لما يلزمه الإنسان من حظر الشرع والدخول في أحكامه (137–138).	الحجر ما يلتزمه الإنسان من حظر الشرع والدخول في أحكامه.
أمّا اللبّ فهو الذي قد خلص من عوارض الشبه وترشّح لاستفادة الحقائق من دون الفزع إلى الحواسّ (138).	اللبّ ما يخلص به من عوارض الشبه ويترشّح لاستفادة الحقائق من دون المفزع إلى الحواسّ.
العلم إدراك الشيء بحقيقته. وهو ضربان: أحدهما حصول صور المعلومات في النفس. والثاني حكم النفس على الشيء بوجود شيء له هو موجود أو نفي شيء عنه هو غير موجود له (140).	العلم إدراك الشيء بحقيقته. ويطلق على حصول صور المعلومات في النفس. وقيل: حكم النفس على الشيء بوجود شيء له هو موجود أو نفي شيء عنه هو غير موجود له.

(*cont.*)

al-Dharīʿa ilā makārim al-sharīʿa	*Maqālīd al-ʿulūm*
أمّا الدراية فالمعرفة المدركة بضرب من الحيل، وهو تقديم المقدّمة وإجالة الخاطر واستعمال الرويّة (141).	الدراية المعرفة المدركة بضرب من الحيل، وهو تقديم المقدّمة وإجالة الخاطر واستعمال الرويّة.
أمّا الفضيلة فاسم لما يحصل به الإنسان ميّزة على غيره، وهي اسم لما يتوصّل به إلى السعادة (109).	الفضيلة ما يحصّل به الإنسان مزيّة على غيره متوصّلاً به إلى السعادة.
وأمّا النافع فهو ما يعين على بلوغ الفضيلة والسعادة والخير (109).	النافع ما يعين على بلوغ الفضيلة والسعادة والخير.
أمّا الحسب فقد يقال فيما يختصّ الإنسان به فيعدّه من مآثره (118).	الحسب ما يختصّ به الإنسان فيعدّه من مآثره.
الإحسان تحرّي الحسنى في الإيمان والإسلام (160).	الإحسان تحرّي الحسنى في الإيمان والإسلام.
الإخلاص أن يقصد الإنسان فيما يفعله وجه الله تعالى متعرّياً عن الالتفات إلى غيره (161).	الإخلاص أن يقصد بما يفعله وجه الله متعرّياً عن الالتفات إلى غيره.
البرّ السعة في علم الحقّ وفعل الخير (160).	البرّ السعة في علم الحقّ وفعل الخير.
البذاء الكلام القبيح (201).	البذاء الكلام القبيح.
الرفث فواحش الكلام في باب النكاح وأوصاف النساء (201).	الرفث فواحش الكلام في النكاح وأوصاف النساء.
الوقاحة [...] وحقيقتها لجاجة النفس في تعاطي القبيح. (207).	الوقاحة لجاج النفس في تعاطي القبيح.
المشاورة [...] وهي استنباط المرء الرأي من الغير فيما يعرض له من مشكلات الأمور، ويكون ذلك في الأمور الجزئية التي [...] (210).	المشاورة استنباط الرأي من الغير فيما يعرض من مشكلات الأمور الجزئية.

(*cont.*)

al-Dharīʿa ilā makārim al-sharīʿa	*Maqālīd al-ʿulūm*
النصح [...] هو إخلاص المحبّة لغيره في إظهار ما فيه صلاحه (211).	النصيحة إخلاص المحبّة لغيره في إظهار ما فيه صلاحه.
التواضع [...] وهو رضا الإنسان بمنزلة دون ما يستحقّه فضله ومنزلته (213).	التواضع [...] وقيل: الرضا بمنزلة دون ما يستحقّه فضله ومنزلته.
الغيرة ثوران الغضب حماية على أكرم الحرم (244).	الغيرة ثوران الغضب حماية عن الحرم.
العدو هو الذي يتحرّي اغتيال الآخر ويضادّه فيما يؤدّي إلى مصالحه (261).	العداوة تحرّي اغتيال الغير ومضادّته فيما يؤدّي إلى مصالحه.

Chapter 21: Sufism (*al-taṣawwuf*)

The two main sources for this chapter are Kalābādhī's *Taʿarruf* and Qushayrī's *Risāla*. The order of the terms follows that of the *Risāla*. The author used this work from *bāb* 4 onwards. *al-Taʿarruf* was used as a complementary source from which alternative definitions were taken. There are, however, some definitions which are solely based on *al-Taʿarruf*. The table below provides some examples. Although these two treatises were the major sources for this chapter, the chapter contains a few terms and definitions, which cannot be found in either of the two sources (for ex. *al-qabḍ*, *al-basṭ*, and *al-hiyba*).

al-Taʿarruf / *al-Risāla*	*Maqālīd al-ʿulūm*
اعلم أنّ علوم الصوفية علوم الأحوال والأحوال مواريث الأعمال ولا يرث الأحوال إلّا من صحّح الأعمال (التعرّف، 58).	التصوّف هو العلم بالأحوال الموروثة من تصحيح الأعمال ظاهراً وباطناً.

(*cont.*)

al-Taʿarruf/al-Risāla	*Maqālīd al-ʿulūm*
الوقت حادث متوهّم علّق حصوله على حادث متحقّق (الرسالة، 130).	الوقت حادث متوهّم علّق حصوله على حادث متحقّق.
أشار القوم بالفناء إلى سقوط الأوصاف المذمومة (الرسالة، 148). فالفناء هو أن يفني عنه الحظوظ فلا يكون له في شيء من ذلك حظّ ويسقط عنه التمييز فناءاً عن الأشياء كلّها شغلاً بما فني به (التعرّف، 92).	الفناء سقوط الأوصاف المذمومة. وقيل: أن يفني عنه الحظوظ فلا يكون له في شيء حظّ ويسقط عنه التمييز فناءاً عن الأشياء كلّها شغلاً بمن فني فيه.
التفرقة التي هي عقيب الجمع هو أن يفرّق بين العبد وبين هرمومه في حظوظه وبين طلب مرافقه وملاذّه. [...] معناه أنّ التقرّب إليه بالأعمال تفرقة وإذا شاهدوه مقرّباً لهم فهو الجمع (التعرّف، 88–89).	التفرقة أن يفرّق بين العبد وبين همومه في حظوظه بطلب مرافق وملاذّه. وقيل: التقرّب إليه بالأعمال المتفرّقة فإذا شاهدوه مقرّباً إليهم فهو الجمع.
الأنس [...] ارتفاع الحشمة مع وجود الهيبة. [...] أن يستأنس بالأذكار فيغيب عن رؤية الأغيار. [...] انبساط المحبّ إلى المحبوب (التعرّف، 76–77).	الأنس معنىً يعرض لاستشعار لطف المشاهد. وقيل: ارتفاع الحشمة مع وجود الهيبة. وقيل: أن يستأنس بالأذكار فيغيب عن رؤية الأغيار. وقيل: انبساط المحبّ إلى المحبوب.
التواجد استدعاء الوجد بضرب الاختيار. (الرسالة، 139). التواجد ظهور ما [يجد] في باطنه على ظاهره (التعرّف، 82).	التواجد استدعاء الوجد بضرب من الاختيار. وقيل: ظهور ما يجد في باطنه على ظاهره.

Bibliography

Abdel Haleem, M.A.S. 'Early Islamic Theological and Juristic Terminology: *Kitāb al-ḥudūd fī l-uṣūl*, by Ibn Fūrak', *Bulletin of the School of Oriental and African Studies* 54 1 (1991): 5–41 (5–11).

Āmulī, Sayyid Ḥaydar. *Tafsīr al-Muḥīṭ al-aʿẓam wa-l-baḥr al-khaḍam fī taʾwīl kitāb Allāh al-ʿazīz al-muḥkam*, ed. Sayyid Muḥsin al-Mūsawī al-Tabrīzī, 4 vols., Qom: Muʾassasa-yi Farhangī u Nashr-i Nūrun ʿAlā Nūr, 1380 Sh./2001–2002.

Āmulī, Shams al-Dīn. *Nafāʾis al-funūn fī ʿarāʾis al-ʿuyūn*, ed. Abū l-Ḥasan Shaʿrānī, 3 vols., Tehran: Intishārāt-i Islāmiyya, 1377–1379/1957–59.

Anṣārī Shīrāzī, ʿAlī b. Ḥusayn. *Ikhtiyārāt-i badīʿī*, ed. Muḥammad Taqī Mīr, Tehran: Shirkat-i Dārūyī Pakhsh-i Rāzī, 1371 Sh./1992–1993.

Āqā Buzurg Ṭihrānī. *al-Dharīʿa ilā taṣānīf al-Shīʿa*, 25 vols., Beirut: Dār al-Aḍwāʾ, 1403–1406/1983–1986.

Astarābādī, Raḍī al-Dīn. *Sharḥ al-Raḍī ʿalā l-Kāfiya*, ed. Yūsuf Ḥasan ʿUmar, Tehran: Muʾassasat al-Ṣādiq. 1384/1964–1965.

al-Baghawī, Abū Muḥammad al-Ḥusayn. *Tafsīr al-Baghawī*, ed. Muḥammad ʿAbdullāh al-Namir, ʿUthmān Jumʿa al-Ḍaymariyya, and Sulaymān Muslim al-Ḥarash, Riyadh: Dār Ṭayyiba, 1409/1988–1989.

Basharī, Jawād. 'Barrasī-i awḍāʿ-i adabī-i Fārs dar sada-yi hashtum u manābiʿ-i īn barrasī', *Āyina-yi Pazhūhish* 17 1 (1385 Sh./2006): 69–80.

Bashir, Shahrzad. *Fazlallah Astarabadi and the Hurufis*, Oxford: One World, 2005.

Bayḍāwī, Nāṣir al-Dīn. *Minhāj al-wuṣūl ilā ʿilm al-uṣūl*, ed. Shaʿbān Muḥammad Ismāʿīl, Beirut: Dār al-Ḥazm, 1429/2008.

Bayḍāwī, Nāṣir al-Dīn. *Ṭawāliʿ al-anwār min maṭāliʿ al-anḍār*, ed. ʿAbbās Sulaymān. Dār al-Jīl, Beirut, 1411/1991.

Bosworth, C.E. 'A Pioneer Arabic Encyclopedia of the Sciences: al-Khwārazmī's Keys of the Sciences', *ISIS* 14 (1964): 97–111.

Bosworth, C.E. 'Some New Manuscripts of al-Khwarazmi's *Mafātīh al-ʿulūm*', *Journal of Semitic Studies* 9 2 (1964): 341–345.

Brentjes, Sonja. 'Courtly Patronage of the Ancient Sciences in Post-Classical Islamic Societies', *al-Qanṭara* 29 2 (2008): 403–436.

Brockelmann, Carl. *Geschichte der arabischen Litteratur*, 2 vols., Leiden: 1943–1949.

Brookshaw, Dominic Parviz. 'Jahān-Malek Ḵātūn', *Encyclopædia Iranica*, vol. 14, pp. 383–385. *Iranica*, vol. 14, pp. 383–385.

Chulūngar, ʿAlī. 'Taṣḥīḥ u taḥshiya-yi *Kunūz al-wadīʿa min rumūz al-Dharīʿa*', PhD Dissertation: University of Isfahan, Isfahan, 1392 Sh./2013.

Dāmghānī, Iftikhār al-Dīn and Ḥasan Dhulfaqārī (ed.). 'Rawshanāyī-nāma', in Īraj Afshār, ed., *Nāmwāra-yi duktur Maḥmūd Afshār*, 16 vols., Tehran: Bunyād-i Mawqūfāt-i Duktur Maḥmūd Afshār, 1373 Sh./1994, vol. 8, pp. 4300–4305.

de Bruin, J.T.P. "Imād Faḳīh Kirmānī', *Encyclopaedia of Islam, 2nd Ed.*, suppl., pp. 414–416.

Eichner, Heidrun. *Towards the Construction of Islamic Orthodoxy: Philosophy in the Post-Avicennian Period and Islamic Theology as Literary Traditions*, Habilitationsschrift, Martin-Luther-Universität Halle-Wittenberg 2009.

Faqīh Kirmānī, 'Imād al-Dīn. *Dīwān-i 'Imād-i Faqīh-i Kirmānī*, ed. Rukn al-Dīn Humāyūn, Tehran: Ibn Sīnā, 1348 Sh./1969.

Farmer, Henry George. 'The Science of Music in the *Mafātīḥ al-'ulūm*', *Transactions Glasgow University Oriental Society*, XVII (1957–1958): 1–9.

Farmer, Henry George. *The Sources of Arabian Music: An Annotated Bibliography of Arabic Manuscripts Which Deal with the Theory, Practice, and History of Arabian Music*, Bearsden: Issued privately by the author, 1940.

Farmer, Henry George. "Ūd', *Encyclopaedia of Islam*, vol. 8, Leiden: Brill, 1938, pp. 985–988.

Ghanī, Qāsim. *Tārīkh-i 'aṣr-i Ḥāfiẓ: Baḥth dar āthār u afkār u aḥwāl-i Ḥāfiẓ*, Tehran: Zawwār, 1383 Sh./2004.

Ḥājjī Khalīfa. *Kashf al-ẓunūn 'an asāmī al-kutub wa-l-funūn*, 2 vols., Tehran, 1387/1967.

Hilālī Qazwīnī, 'Alā' al-Dīn. *Manāhij al-ṭālibīn fī ma'ārif al-ṣādiqīn*, ed. Aḥmad Bihnāmī, Tehran: Kitābkhāna, Mūzi u Markaz-i Asnād-i Markaz-i Shūrā-yi Islāmī, 1397 Sh./2018.

Ibn al-Ḥājib, Jamāl al-Dīn 'Uthmān. *al-Kāfiya fī 'ilm al-naḥw wa-l-shāfiya fī 'ilmay al-taṣrīf wa-l-khaṭṭ*, ed. Ṣāliḥ 'Abd al-'Aẓīm al-Shā'ir, Cairo: Maktabat al-Ādāb, 1431/2010.

Ibn Jamā'a, Badr al-Dīn. *al-Manhal al-rawī fī mukhtaṣar 'ulūm al-ḥadīth al-nabawī*, ed. Muḥyī al-Dīn 'Abd al-Raḥmān Ramaḍān, Damascus: Dār al-Fikr, 1406/1985–1986.

Ibn al-Nafīs, 'Alā' al-Dīn al-Qurashī al-Dimashqī. *al-Ṣaydaliyya al-mujarraba*, ed. Muḥsin 'Aqīl, Beirut: Dār al-Maḥajja al-Bayḍā', 1423/2002.

Ibn Nāẓim al-Dimashqī. *al-Miṣbāḥ fī l-ma'ānī wa-l-bayān wa-l-badī'*, ed. Ḥusnī 'Abd al-Jalīl Yūsuf. Cairo: Maktabat al-Ādāb, 1409/1989.

Ibn Ṣalāḥ al-Shahrazūrī, Abū 'Amr 'Uthmān. *'Ulūm al-ḥadīth*, ed. Nūr al-Dīn 'Atar, Damascus: Dār al-Fikr, 1406/1986.

Ibn Sarrāj, Abū Bakr al-Shantarīnī. *Kitāb al-kāfī fī 'ilm al-qawāfī*, ed. D. 'Alā' Muḥammad Ra'fat, Cairo: Dār al-Ṭalā'i', 2003.

Ibn Sīnā. *al-Qānūn fī l-ṭibb*, ed. Muḥammad Amīn al-Ḍanāwī. Beirut: Dār al-Kutub al-'Ilmiyya, 1420/1999.

Ibn Sīnā. 'Risāla al-Ḥudūd', in 'Abd al-Amīr al-A'ṣam, ed., *Rasā'il manṭiqiyya li-l-falāsifa al-'Arab*, Beirut: Dār al-Manāhil. 1428/2008, pp. 111–153.

Ibn Sīnā. *'Uyūn al-ḥikma*, ed. 'Abd al-Raḥmān Badawī, 2nd Ed., Beirut: Dār al-Qalam, 1980.

Ingenito, Domenico. 'Jahan Malik Khatun: Gender, Canon, and Persona in the Poems of a Premodern Persian Princes'. In Alireza Korangi, Hanadi Al-Samman, Michael

C. Beard, ed., *The Beloved in Middle Eastern Literature: The Culture of Love and Languishing*, London and New York: I.B. Tauris, 2018, pp. 177–212, 330–337.

Īmānī, Bihrūz. 'Durarī az daryā-yi mutūn (4)', *Payām-i Bahāristān* 6 23 (1393 Sh./2014): 349–356.

al-Isfarā'inī, Tāj al-Dīn. *al-Lubāb fī 'ilm al-i'rāb*, ed. Shawqī al-Ma'arrī, Beirut: Maktabat Lubnān Nāshirūn, 1996.

Jahān-Malik Khātūn. *Dīwān-i kāmil-i Jahān-Malik Khātūn, qarn-i hashtum-i hijrī*. Ed. Pūrāndukht Kāshānī Rād and K. Aḥmadnizhād. Tehran: Zawwār, 1374 Sh./1995.

al-Jārbardī, Fakhr al-Dīn. 'Sharḥ al-Shāfiya', in *Majmū'a al-Shāfiya fī 'ilmay al-taṣrīf wa-l-khaṭṭ*, Muḥammad 'Abd al-Salām Shāhīn, Beirut: Dār al-Kutub al-'Ilmiyya, 1971.

al-Jurjānī, Sayyid Sharīf. *Kitāb al-Ta'rīfāt*, ed. Ibrāhīm al-Abyārī. Cairo: Dār al-Rayyān li-l-Turāth, 1403/1982–1983.

al-Jurjānī, Sayyid Sharīf and Maḥmūd Yazdī Muṭlaq Fāḍil (ed.). 'Sayrī dar aḥwāl u āthār-i Mīr Sayyid Sharīf-i Gurgānī', in Jūyā Jahānbakhsh and Bahā' al-Dīn Khurramshāhī, ed., *Muḥaqqiq-Nāma: Maqālāt-i taqdīm-shuda bi ustād duktur Mahdī Muḥaqqiq*, 2 vols., Tehran: Anjuman-i Āthār u Mafākhir-i Farhangī, 1380 Sh./2001–2002, vol. 2, pp. 1389–1447.

al-Kalābādhī, Abū Bakr Muḥammad b. Isḥāq. *al-Ta'arruf li-madhhab ahl al-taṣawwuf*, ed. Aḥmad Shams al-Dīn, Beirut: Dār al-'Ilm, 1422/2001–2002.

Karabela, Mehmet Kadri. 'The Development of Dialectic and Argumentation Theory in Post-Classical Islamic Intellectual History', PhD Dissertation, McGill University, Montreal, 2010.

al-Khaṭīb al-Qazwīnī, Jalāl al-Dīn. *al-Īḍāḥ fī 'ulūm al-balāgha: al-ma'ānī wa-l-bayān wa-l-badī'*, Beirut: Dār al-Kutb al-'Ilmiyya, 1424/2003.

al-Khaṭīb al-Tabrīzī, Yaḥyā b. 'Alī. *al-Kāfī fī l-'arūḍ wa-l-qawāfī*, ed. al-Ḥassānī Ḥasan 'Abdullāh, Cairo: Maktabat al-Khānajī, 1994/1415.

al-Kīlānī, Quṭb al-Dīn. *Sharḥ al-Risāla fī Ādāb al-baḥth*, MS Majlis-i Shūrā-yi Islāmī 10361/6, pp. 68–119.

al-Kindī. 'Risāla al-Ḥudūd', in 'Abd al-Amīr al-A'ṣam, ed., *Rasā'il al-manṭiqiyya li-l-falāsifa al-'Arab*, Beirut: Dār al-Manāhil, 1428/2008, pp. 59–79.

Khʷārazmī, Abū 'Abdullāh Muḥammad. *Liber Mafâtîh al-olûm: explicans vocabula technica scientiarum tam Arabum quam peregrinorum*, ed. Gerlof van Vloten, Leiden: Brill, 1895.

Kutubī, Maḥmūd. *Tārīkh-i Āl-i Muẓaffar*, ed. 'Abd al-Ḥusayn Nawā'ī, Tehran: Intishārāt-i Amīr Kabīr, 1364 Sh./1985.

Mubārakshāh. *Tarjuma-yi sharḥ-i Mubārakshāh-i Bukhārī bar Adwār-i Urmawī dar mūsīqī*, ed. and trans. into Persian by Sayyid 'Abdullāh Anwār, Tehran: Farhangistān-i Hunar, 1392 Sh./2013.

Nawīdī Malāṭī, 'Alī. 'Nasīm al-rabī'-i Ṣūfī-i Samarqandī: Tarjuma-yi qadīm-i Rabī' al-abrār-i Samarqandī', *Nāma-yi Bahāristān* 17 (1389 Sh./2010): 167–180.

Pourjavady, Amir Hosein. 'Jurjānī, 'Alī b. Muḥammad: Mūsīqī', *Dā'irat al-ma'ārif-i buzurg-i Islāmī*, ed. Kāẓim Mūsawī Bujnūrdī. Tehran: Markaz-i Dā'irat al-Ma'ārif-i Buzurg-i Islāmī, 1388 Sh./2009, vol. 17, pp. 714–715.

Pourjavady, Reza. 'The Legacy of Aḍud al-Dīn al-Ījī: His Works and His Students', in Ayman Shihadeh and Jan Thiele, ed., *Philosophical Theology in Islam: The Later Ash'arism East and West*, Leiden and Boston: Brill, 2020, pp. 337–370.

Qazwīnī, Muḥammad. *Yāddāshthā-yi Qazwīnī*, ed. Īraj Afshār, Tehran: Intishārāt-i Dānishgāh-i Tihrān, 1388 Sh./2009.

al-Qushayrī, Abū l-Qāsim. *al-Risāla al-Qushayriyya*, ed. 'Abd al-Ḥamīd Maḥmūd and Maḥmūd b. Sharīf, Cairo: Dār al-Sha'b, 1409/1989.

al-Rāghib al-Iṣfahānī, Abū l-Qāsim. *al-Dharī'a ilā makārim al-sharī'a*, ed. Abū Yazīd Abū Zayd al-'Ajamī, Cairo: Dār al-Islām, 1428/2007.

al-Rāzī, Quṭb al-Dīn. *Taḥrīr al-qawā'id al-manṭiqiyya fī sharḥ al-Risāla al-Shamsiyya*, ed. Muḥsin Bīdārfar, Qom: Intishārāt-i Bīdār, 1386 Sh./2007.

Rieu, Charles. *Supplement to the Catalogue of the Arabic Manuscripts in the British Museum*, London: The British Museum, 1894.

Rosenfeld, Boris A. and Ekmeleddin İhsanoğlu. *Mathematicians, Astronomers & Other Scholars of Islamic Civilization and Their Works (7th–19th c.)*, Istanbul: Research Center for Islamic History, Art and Culture, 2003.

al-Sakkākī, Sirāj al-Dīn. *Miftāḥ al-'ulūm*, ed. Nu'aym Zarzūr, Beirut: Dār al-Kutub al-'Ilmiyya, 1987.

Ṣā'in Hirawī, Rukn al-Dīn. *Dīwān-i Ṣā'in-i Hirawī*, Patna: Mu'assasa-yi Taḥqīqāt u Tatabbu'āt dar Zabān u Adab-i 'Arabī u Fārsī, 1959.

al-Samarqandī, Shams al-Dīn. *Risāla fī ādāb al-baḥth*. MS Majlis-i Shūrā-yi Islāmī 9006/1, pp. 3–8.

al-Samarqandī, Shams al-Dīn. *al-Ṣaḥā'if al-ilāhiyya*, ed. Aḥmad 'Abd al-Raḥmān al-Sharīf. Kuwait City: al-Fallāḥ, 1985.

Samarqandī, Shams al-Dīn. *Science of the Cosmos and the Soul: 'Ilm al-'āfāq wa al-'anfus*, ed. Gholamreza Dadkhah, California, Costa Mesa: Mazda Publishers, 2014.

Ṣarrāf Yazdī, 'Aḍud al-Dīn. *Dīwān-i 'Aḍud*, ed. 'Alī Riḍā Qūjazāda, Tehran: Kitābkhāna, Mūzi u Markaz-i Asnād-i Majlis-i Shūrā-yi Islāmī, 1389 Sh./2010.

Sāwajī, Salmān. *Dīwān*-i *Salmān-i Sāwajī*, ed. Manṣūr Mushfiq with an introduction by Taqī Tafaḍḍulī, Tehran: Ṣafī 'Alī-Shāh. 1367 Sh./1988–1989.

Seidel, E. 'Die Medizin in Kitâb Mafâtîḥ al-'Ulûm', *Sitzungsberichte der physikalisch-medizinischen Sozietät in Erlangen* XLVII (1915): 1–79.

Shakībāniyā, Mahdī and Reza Pourjavady. 'Kitābshināsī-i Sayyid Sharīf-i Jurjānī', *Ma'ārif* 19 3 (2003): 134–192.

Shamsī, Ghulām-Riḍā. 'Muqaddima, taṣḥīḥ u taḥlīl-i kitāb-i *Kunūz al-wadī'a min rumūz al-Dharī'a*', PhD Dissertation: University of Kurdistan, Sanandaj, 1395 Sh./2016–2017.

Shiloah, Amnon. *The Theory of Music in Arabic Writings (C. 900–1900): Descriptive Catalogue of Manuscripts in Libraries of Europe and U.S.A.*, Munich: RISMB, 1979.

al-Shīrāzī, Muʿīn al-Dīn Junayd. *Shadd al-izār fī ḥaṭṭ al-awzār ʿan zuwwār al-mazār*, ed. Muḥammad Qazwīnī and ʿAbbās Iqbāl, Tehran: Chāpkhāna-yi Majlis, 1328 Sh./1949.

al-Suyūṭī, Jalāl al-Dīn. *Bughyat al-wuʿāh fī ṭabaqāt al-lughawiyyīn wa-l-nuḥāh*, Cairo: Maṭbaʿat ʿĪsā al-Bābī al-Ḥalabī, 1964.

al-Suyūṭī, Jalāl al-Dīn (attrib.). *Maqālīd al-ʿulūm fī l-ḥudūd wa-l-rusūm*, ed. Muḥammad Ibrāhīm ʿUbāda, Cairo: Maktabat al-ādāb, 1424/2004.

Ṭabīb Shīrāzī, Jalāl al-Dīn. *Dīwān-i Jalāl-i Ṭabīb-i Shīrāzī*, ed. Nasrollah Pourjavady, Tehran: Farhangistān-i Zabān u Adab-i Fārsī, 1389 Sh./2010.

al-Ṭayyibī, Sharaf al-Dīn. *al-Khulāṣa fī l-ḥadīth*, ed. Abū ʿĀṣim al-Shawāmī al-Atharī, Cairo: al-Maktaba al-Islāmiyya, 1430/2009.

al-Ṭayyibī, Sharaf al-Dīn. *al-Tibyān fī l-bayān*, ed. Yaḥyā Murād, Beirut: Dār al-Kutub al-ʿIlmiyya, 1425/2004.

van Ess, Josef. *Die Träume der Schulweisheit: Leben und Werk des ʿAlī b. Muḥammad al-Ǧurǧānī (gest. 816/1413)*, Wiesbaden: Harrassowitz Verlag, 2013.

van Ess, Josef. 'Ḥaydar-i Āmulī', *Encyclopaedia of Islam, 2nd ed.*, Supplement volume XII, pp. 363–365.

van Ess, Josef. 'Jorjāni, Zayn-al-Din Abu'l-Hasan ʿAli', *Encyclopædia Iranica*, vol. 15/1, 2009, pp. 21–29.

Wiedemann, E. 'Beiträgen zur Geschichte der Naturwissenschaft X. Zur Technik bei den Arabern', *Sitzungsberichte der physikalisch-medizinischen Sozietät in Erlangen* XXXVIII (1906): 307–357.

Wiedemann, E. 'Beiträgen zur Geschichte der Naturwissenschaft XIV. Über die Geometrie und Arithmetik nach den Mafâtiḥ al-ʿUlûm', *Sitzungsberichte der physikalisch-medizinischen Sozietät in Erlangen* XL (1908): 1–29.

Wiedemann, E. 'Beiträgen zur Geschichte der Naturwissenschaft XXII. Stucke aus den Mafâtîḥ al-ʿUlûm', *Sitzungsberichte der physikalisch-medizinischen Sozietät in Erlangen* XLII (1910): 303–322.

Wiedemann, E. and W. Müller. 'Beiträgen zur Geschichte der Naturwissenschaft LXVI. Zur Geschichte der Musik', *Sitzungsberichte der physikalisch-medizinischen Sozietät in Erlangen* LIV–LV (1922–1923): 7–16.

Yazdī, Muʿīn al-Dīn. *Mawāhib-i ilāhī dar tārīkh-i Āl-i Muẓaffar*, ed. Saʿīd Nafīsī, vol. 1, Tehran: Kitābkhāna u Chāpkhāna-yi Iqbāl, 1326 Sh./1947.

Zākānī, ʿUbayd. *Kulliyyāt-i ʿUbayd-i Zākānī*, ed. ʿAbbās Iqbāl and Parwīz Atābakī, Tehran: Iqbāl, 1340 Sh./1961.

Zaydan, Jurji. *Tārīkh ādāb al-lugha al-ʿarabiyya*, Cairo: Dār al-Hilāl, 1957.

al-Ziriklī, Khayr al-Dīn. *al-Aʿlām*, Beirut: Dār al-ʿIlm Li-l-Malāyīn, 2002.

مقاليد العلوم

في الحدود والرسوم

لمؤلّف مجهول عاش في القرن الثامن الهجري

حقّقه وقدّم له

رضا پورجوادي
وغلامرضا دادخواه

TUTA SUB AEGIDE PALLAS · 1683 ·
بريل

الناشر

دار بريل للنشر في ليدن المحروسة وبوسطن

فهرس المحتويات

[خطبة الكتاب]

بسم الله الرحمن الرحيم

الحمد لله الذي ليس لماهيته حدّ فيعرب عنه لسان، ولم تدخل[1] معرفة ذاته تحت رسم الحدوث[2] والإمكان، فالتامّ والناقص بكمال عظمته وجلال عزّته معترفان، والعالم والجاهل من تيّار بحار رحمته وزلال[3] إفضال نعمته مغترفان. وصلّى الله على مبين[4] حدود الحرام والحلال، ومعفي رسوم البدع والضلال، محمد المرسل إلى خلقه للتعريف والتمييز[5]، المنزل فيه ﴿لَقَدْ جاءَكُمْ رَسُولٌ مِنْ أَنفُسِكُمْ عَزِيزٌ﴾ [6]، وعلى آله الموفين بمواثيق العهود، الواقفين[7] على مدارك الحدود، صلاة جامعة لكلّ إفضال وإكرام، مانعة عن الانفصال[8] والانفصام[9]، تطّرد أدوارها مع تعاقب الأزمان والدهور، وتنعكس[10] أنوارها على[11] صفحات السنين والشهور، وسلّم تسليماً.

أمّا بعد، فإنّ معرفة المواضعات والمصطلحات من أوائل الصناعات وأهمّ المهمّات. والطالب الذهن الأديب، الراغب الفطن اللبيب، متى فرغ عن حفظ اللغة واستحضارها[12]، وضبط أنواع مفرداتها[13] واستظهارها[14]، لا بدّ وأن يكون بمصطلحات أهل كلّ فنّ خبيراً، وبمواضعات كلّ طبقة من العلماء بصيراً، ليحيط به إحاطة أوّلية تكون له عوناً على التحصيل، ويطّلع على مقاصدهم إجمالاً قبل التفصيل. حتّى إذا أراد استحضار[15] مسائلها وأحكامها، والوقوف على جميع أنواعها وأقسامها، سهل عليه ما يريده وحصل به إتقانه وتسديده. فلم يتلعثم في بيان جواب، ولم يتتعتع في دراسة علم وكتاب. فلكلّ طائفة من العلماء كلمات فيما بينهم متعارفة، لا يفهم مرادهم منها إلّا من بلغ مبلغ قصدهم أو شارفه.[16]

[1] تدخل: يدخل، آ ب س. [2] الحدوث: الحدث، ب. [3] زلال: ذلال، ب. [4] مبين: مبيّن، س. [5] التمييز: التميين، آ؛ التميّز، ب. [6] التوبة، 128. [7] الواقفين: الوافقين، آ. [8] الانفصال: الانفضال، آ. [9] الانفصام: الانقصال، آ. [10] تنعكس: ينعكس، ب. [11] على: -، ب. [12] استحضارها: استحضرها، ب. [13] مفرداتها: مفرداته، آ ب س. [14] استظهارها: استظهرها، ب. [15] استحضار: استحصال، ب. [16] شارفه: شارقه، آ؛ شارقة، س.

 | DOI:10.1163/9789004423367_007

وربّ كلمة لم يتجاوز فهم اللغوي عن حقيقتها، ولم تعرف[1] متصرّفات الأقوام في طريقتها. فإنّ لفظة الواجب مثلاً عند الفقيه غير ما عند الأصولي، ولفظة السند عند المحدّث غير ما عند الجدلي. وكذا [لفظة] الحال عند النحوي غير ما عند الصوفي[2]، والوتد عند المنجّم غير ما عند العروضي، والرجعة في عرف الفقيه غير ما في عرف المتكلّم، وهي في عرف الكتّاب غير ما في[3] عرف المنجّم، إلى غير ذلك مما ستقف عليه في كلّ باب من أبواب هذا الكتاب، إذ لا مشاحّة في الاصطلاحات ولا خصومة في الشهوات.

هذا، وكم من طالب متفنّن قد برّز على أقرانه وافتخر بنباهته وعلوّ شأنه. ثم إذا سئل في بعض المباحث عن تعريف مسّ إليه الاحتياج، زال عنه الطلاقة والابتهاج، وآل أمر ذلاقته إلى الارتتاج، فوقع في داء عضال، وحجّ عليه الخصم عند الجدال، واعتراه عقيب دعواه حمرة[4] الخجل وصفرة الوجل. فمثله كالعامل الجائر[5] بعد ما عزل. وما أحسن قول الشاعر في الزمن الغابر:

بيت[6]

من تحلّى[7] بغير ما هو فيه فَضَحَتْهُ شواهد الامتحان[8]

ثمّ إنّ كتابي هذا المترجم بمقاليد العلوم في الحدود والرسوم، جامع لمصطلحات أكثر الفنون، حاوٍ لمقدمات الأقسام على ما أدّت إليه الظنون، كفيل أن يكون لكلّ فنّ مدخلاً كافياً، حقيق أن يعدّه المبتدئ ذخراً[9] وافياً. وقد كنت أكتبه أشتاتاً في أوائل الشباب، حال مدارسة العلوم والآداب، والاندراج في زمرة الطلاب، واصطكاك الركب لاختيار النحب بين يدي الأساتذة النحارير والأئمّة الأعلام المشاهير، رحم الله الماضين وأدام عمر الباقين. فمنها ما لقفت عن فلق أفواههم بين حلق[10] المدارس، وثقفت عن فرائد محاوراتهم في الأندية[11] والمجالس. ومنها ما نقلت عن كتب تداولوها، وأسئلة وأجوبة تقاولوها، وكلّها تعريفات مستقيمة وحدود أو رسوم قويمة. فرغّبني بعض إخواني في الدين وأحد خلّاني من أهل اليقين أن أجمعها من ظهور الدفاتر[12] وبطون الأوراق، قبل أن يتبدّل

[1] تعرف: يعرف، آ س. [2] الصوفي: الصرفي، آ. [3] في: -، ب. [4] حمرة: خمرة، آ ب. [5] الجائر: الجابر، آ س. [6] بيت: شعر، آ س. [7] تحلّى: تجلّى، آ. [8] البيت مذكور في العمدة في صناعة الشعر والنقض لابن الرشيق القيرواني، حقّقه النبوي عبد الواحد شعلان (القاهرة: مكتبة الخانجي، 1420/2000)، الجزء الثاني، ص 991. [9] ذخراً: زخراً، آ س. [10] حلق: خلق، آ. [11] الأندية: الأبدية، آ. [12] الدفاتر: الدقائق، آ.

الزمان ويتبدّد سلك الاتّفاق. فبوّبتها على أحد وعشرين باباً قصيرة عن طويلة، انتخبتها انتخاباً يشتمل كلّ باب على تعريف علم أوّلاً، ثمّ على مصطلحات ذاك العلم مجملاً. وإن اختلطت فيه علوم العقل والشرع، فلعلّ الوضع عند التأمّل يوافق الطبع. فمن وزنها بميزان العقل، وكفّ لسان الطعن[1] والعذل، سيجد الجامع في ترتيب هذا الكتاب مصيباً، وأنه قد رزق، إن شاء الله، من الحكمة وفصل الخطاب نصيباً.

وحين سجّل القضاء بالإتمام، وعُدْتُ من الافتتاح إلى الاختتام، رجوت أن يتشرّف[2] بمطالعة العلماء الفحول، ويلاحظه الفضلاء بنظر اللطف والقبول، ليجد رواجاً في سوق الأزمان، ويبقى به ذكري بين الإخوان، فصدّرته بذكر ألقاب السلطان الأعظم، الخاقان الأفضل الأعلم، ظلّ الله في أرضه، والقائم بنفله وفرضه، ويده المبسوطة على خلقه، وأمينه المؤتمن[3] على حقّه، مولى ملوك العالمين، وارث علوم الأنبياء والمرسلين، محيي مآثر الخلفاء الراشدين، جلال الحقّ والدنيا والدين، السلطان المطيع المطاع، المدافع على الدين المبين أمتع الدفاع، والساطع نور عدله في سائر الأصقاع[4]، أبي[5] الفوارس شاه شجاع، خلّد الله سلطانه[6]، وأعلى في الدارين شأنه. فهو الجامع المستحضر لأقسام العلوم، المتخصّص به من جميع السلاطين القروم، يحثّ على العلم والفضل، وينقذ[7] عن ورطات الشكّ والجهل، ويحلّ دقائق الأشكال، ويزيل عوارض[8] الإشكال. له سؤالات أحلى[9] من الشهد شيب بماء الغمام، تترك الأذهان مستشرفة[10] لها بين إقدام وإحجام، وجوابات أحسن من الدرّ والعقيان في نحور الغيد الحسان، وأزكى[11] من حركات النسيم بين الورد والريحان. فمجامع العلماء بنشر فوائده ونكاته معمورة، ومحافل[12] الفضلاء بذكر إيراداته وتحقيقاته مغمورة. يتقرّب إلى حضرته العليا كلّ بما عنده، ويعانى في طلب مرضاته وكدَه وكدَّه. فإن نظر إلى هذه البضاعة المزجاة بعين الرضا فهو غاية المسؤول[13] ونهاية المرتجى، وإلّا لم يقع من حسن القبول موقعاً، ولم يجد لشرف المثول موضعاً، فذاك[14] من قلّة استعداد العبد، وقصر الباع[15] عن المرتاد[16] والقصد. وبالله أستعين، وعليه المعوّل، وفيه الرجاء ولديه المؤمّل. ولا حول ولا قوّة إلّا بالله العليّ العظيم.

وقبل الشروع في الأبواب أذكر مقدّمة لهذا الكتاب، والله تعالى ملهم الحقّ والصواب.

---

[1] الطعن: العطن، آ. [2] يتشرّف: تشرف، ب. [3] المؤتمن: المومن، ب. [4] الأصقاع: + الرباع، ب. [5] أبي: أبو، آ س. [6] سلطانه: السلطانه، ب. [7] ينقذ: ينفذ، آ. [8] عوارض: عواض، ب. [9] أحلى: أصلى، ب. [10] مستشرفة: متشرفة، آ، متشرّفة، س. [11] أزكى: ان كى، آ. [12] محافل: محاقد، ب. [13] المسؤول: السؤل، ب. [14] لم يقع من [...] فذك: -، ب. [15] الباع: البان، آ. [16] المرتاد: المرثاد، آ س.

المقدمة: وهي[1] ما يتوقّف عليها المباحث الآتية المقصودة

ومقدّمة كلّ علم ما يتوقّف عليها ذلك العلم. واختلفوا في الفرق بين المقدّمة والمبادئ.

فقيل:[2] لا فرق بينهما.

وقيل: بينهما مباينة كلّية، لأنّ المقدّمة ما يتوقّف عليها الشروع في ذلك العلم، والمبادئ ما يتوقّف عليها البحث عن ذلك العلم.

وقيل: المقدّمة أعمّ من المبادئ مطلقاً. وعرّفوها بأنّ المقدّمة ما يتوقّف عليها المسائل مطلقاً، سواء كان بواسطة أم لا. والمبادئ ما يتوقّف عليها المسائل بلا واسطة.

وقيل: بينهما عموم وخصوص من وجه.

اعلم أنّ معرّف الشيء هو ما يوجب معرفته معرفته. والحقائق إمّا أن تكون بسيطة أو مركّبة. وكلّ منهما إمّا[3] أن يتركّب عنه غيره أو لا يتركّب. فالبسيط الذي لا يتركّب عنه غيره لا يحدّ ولا يحدّ به كالواجب، والذي يتركّب عنه غيره يحدّ به ولا يحدّ[4] كالجوهر.[5] والمركّب الذي لا يتركّب عنه غيره يحدّ ولا يحدّ به كالإنسان، والذي يتركّب عنه غيره يحدّ ويحدّ[6] به كالحيوان. فالحدّ للمركّب، وكذا الرسم التامّ. أمّا الرسم الناقص فيشتملهما.[7]

الحدّ: وهو[8] القول الدالّ على ماهية الشيء.

وقيل: إنّه قول دالّ على ما به الشيء هو[9] هو.

وقيل: هو قول يقوم مقام الاسم في الدلالة على الماهية.

الحدّ التامّ: تعريف الماهية بجميع أجزائها الداخلة.

الحدّ الناقص: تعريفها ببعض أجزائها اللازمة.

الرسم التامّ: أن يذكر الجزء الجنسي، ثمّ يذكر الخاصّة الخارجة[10] عن الماهية مقام الفصل المقوّم لها.

الرسم الناقص: تعريفها بأمور خارجة[11] عنها.

قيل: لو كان للحدّ حدّ لكان لحدّ[12] الحدّ حدّ[13] آخر، وهكذا إلى أن يتسلسل، فيلزم المحال.

---

[1] هي: -، آ. [2] فقيل: قيل، آ. [3] إمّا: -، آ. [4] ولا يحدّ: -، آ. [5] كالجوهر: كالجوهي، آ؛ كالجوهري، ب.
[6] يحدّ: لا يحدّ، آ س. [7] فيشملهما: فيشتملهما، آ ب. [8] هو: وهو، آ ب. [9] هو: + ما، ب. [10] الخارجة: الخارجية، آ ب س. [11] خارجة: خارجية، آ. [12] لحدّ: محدّ، آ. [13] حدّ: حدّاً، آ.

 | DOI:10.1163/9789004423367_008

وأجيب بأنّه إذا حدّ الحدّ المطلق، فقد علم أنّ الحدّ ما هو. فمتى قيل بعده: ما حدّ الحدّ، كان جوابه أن يذكر حدّ الحدّ، ويضاف إلى الحدّ. متى ذكرنا مثلاً حدّ النصف لم نحتج أن نذكر لنصف النصف حدّاً آخر، وهكذا إلى ما لا نهاية له، لأنّا إذا عرفنا أنّ النصف ما هو، علمنا أنّ نصف النصف عبارة عن النصف بشرط إضافته إلى النصف. والله أعلم.

فصل

تعريف الماهية إمّا بنفسها أو بأمور داخلة فيها مقوّمة لها أو بأمور خارجة عنها عارضة لها، وإمّا بتركيب[1] من القسمين.

الأوّل: تعريف الشيء بنفسه محال، لأنّ الوسيلة معلومة[2] قبل المطلوب، فلو جعل الشيء معرّفاً لنفسه لكان ذلك الشيء من حيث إنّه وسيلة يجب أن تكون[3] معرفته متقدّمة، ومن حيث إنّه مطلوب يجب أن تكون معرفته متأخّرة، فيلزم أن يكون العلم به متأخراً عن[4] العلم به، وذلك محال.

الثاني: تعريف الماهية بالأمور[5] الداخلة، إن كان يذكر مجموع أجزائها كان حدّاً تامّاً، لأنّ الماهية إذا كانت مركّبة من مجموع أمور كثيرة، فمتى حصل العلم بجملتها، فقد حصل العلم بتلك الماهية المتركّبة من[6] تلك الأمور، فتكون[7] الصورة الذهنية مطابقة للماهية الخارجية مطابقة لا يمكن الزيادة عليها. فكما أنّ الماهية الخارجية إنّما حصلت من مجموع تلك البسائط، فكذلك العلم بتلك الماهية المركّبة حصل من العلم بمجموع تلك البسائط، فيكون هذا تصوّراً[8] تامّاً كاملاً لا يقبل الزيادة والنقصان، فلذلك سمّوه الحدّ التامّ.

وإن كان التعريف بجزء لازم لتلك الماهية وجوداً وعدماً مما يسمّى فصلاً مقوّماً، فهذا النوع من التعريف، يسمّى حدّاً أو رسماً، فيه اختلاف. قال الإمام رحمه الله: هذا نزاع لفظي، فمن يقول: إنّ الحدّ هو الذي يفيد معرفة تامّة، فهذا النوع عنده ليس بحدّ، ومن يقول: هو الذي يفيد معرفة ذاتية، سواء كانت المعرفة تامّة أو ناقصة، فهذا عنده حدّ.

---

[1] بتركيب: يتركّب، ب. [2] معلومة: معلوم، ب. [3] تكون: يكون، آ ب س. [4] عن: من، آ. [5] بالأمور: بأمور، آ. [6] من: عن، آ س. [7] فتكون: يكون، ب. [8] تصوّراً: التصوّر، ب؛ تصوّر، آ.

وإن كان التعريف بجزء من أجزائها الغير اللازمة لها نفياً وإثباتاً فإمّا أن يكون المعرّف أعمّ من المحدود أو أخصّ منه. فإن كان أعمّ فلا محالة يدخل في التعريف شيء مغاير لذلك المحدود. وإن كان أخصّ فيخرج منه بعض أفراد المحدود[1] فلا يكون جامعاً مانعاً على التقديرين، فلا يطّرد ولا ينعكس فيكون فاسداً.

الثالث: تعريف الماهية بالأمور الخارجة[2] عنها، وشرط هذا المعرّف[3] أن يكون ملازماً للماهية نفياً وإثباتاً، وأن يكون أبين منها وأظهر. وهذا النوع لا يفيد إلّا التمييز[4] فقط، فلذلك سمّوه بالرسم الناقص.

الرابع: تعريف الماهية بمجموع ما يكون داخلاً فيها وما يكون خارجاً عنها، وهذا إمّا أن يذكر[5] جميع مقوّمات تلك الماهية وكلّ ذاتياتها، ثمّ يضمّ إليها الأوصاف الخارجية المعرّفة، فهذا بعيد جدّاً، لأنّ جميع الذاتيات والمقوّمات إذا كانت مذكورة فقد وقع الاستغناء بذكرها عن الأمور الخارجية[6]، وإمّا أن يذكر بعض مقوّماتها وذاتياتها، وهذا على نوعين: إمّا أن يكون بين تلك المقوّمات والخارجيات عموم وخصوص، أو لا يكون. فإن كان الأوّل فإمّا أن يذكر الحدّ الجنسي ثمّ يذكر الخاصّة[7] الخارجية مقام الفصل، فهذا يسمّونه الرسم التامّ، وإمّا أن يذكر الوصف العرضي مقام الجنس ثمّ يردف بالفصل المقوّم. قال الإمام: ولم يتعرّضوا لهذا القسم ولم يذكروه باسم.

والله أعلم.

فصل

قيل: التعريف بمجموع الأجزاء ممتنع، لأنّ مجموع أجزاء الشيء عينه، والجزء إنّما يعرّف إذا عرّف شيئاً من أجزائه. وذلك الجزء إمّا أن يكون هو أو ما هو خارج عنه، والخارج إنما يعرّف إذا علم اختصاصه به. وذلك يتوقّف على معرفته ومعرفة[8] ما يغايره من الأمور الغير المتناهية، وذلك محال.

وأجيب بأنّ الجزء مقدّم على الكلّ بالطبع، والأشياء التي كلّ واحد منها مقدّم على شيء يمتنع[9] أن يكون نفسه معرّفاً[10] به. ومعرّف الشيء ليس بواجب أن يعرّف شيئاً من أجزائه أصلاً، لجواز استغنائها

[1] المحدود: للحدود، آ س. [2] الخارجة: الخارجية، س. [3] المعرّف: تعريف، س. [4] التمييز: التميين، آ. [5] أن يذكر: أن يكون يذكر، آ. [6] الخارجية: الخارجة، ب س. [7] الخاصّة: الخاصّية، آ. [8] معرفة: معرفته، آ. [9] يمتنع: يمنع، ب. [10] معرّفاً: ومعرّفاً، س.

بأسرها. وتعريف الموصوف متوقّف على كون الوصف المعرّف بحيث يلزم من تصوّره تصوّره بعينه، وذلك إنّما يتوقّف حصوله وشموله في نفس الأمر لا على العلم بهما.

قال القاضي رحمه الله: وهو ضعيف، لأنّ تقدّم كلّ واحد لا يقتضي تقدّم الكلّ من حيث هو كلّ ومجموع ليدلّ على المغايرة. ولو كانت الأجزاء بأسرها حتّى الصوريّ[1] معلومة، كانت الماهية معلومة وإلّا لم يفد[2] التحديد. ولو استلزم الخارجي تصوّره تصوّره، فإن كان متصوّراً كان الملزوم متصوّراً، فاستغنى عن التعريف؛ وإن لم يكن[3] متصوّراً امتنع التعريف.

بل الجواب الحقّ أنّ الأجزاء على انفرادها معلومة، والتحديد استحضارها مجموعة بحيث يحصل في الذهن صورة مطابقة للمحدود. فكذا الرسم إذا كان مركّباً. وأمّا المفرد[4] فلا يفيد.

قيل: إن كان المطلوب مشعوراً امتنع تحصيله، وإن لم يكن مشعوراً به امتنع طلبه.

وأجيب بأنّ توجّه الطلب نحو الشيء المشعور به ببعض اعتباراته فلا يكون ممتنعاً. والله أعلم.

[1] الصوريّ: التصوريّ، س. [2] لم يفد: لم يفد، آ؛ لم يفيد، ب. [3] إن لم يكن: إن يكون، آ س. [4] المفرد: مفرد، ب.

الباب الأوّل: في التفسير

التفسير: هو الكلام في أسباب نزول الآية وشأنها وقصّتها وأحكامها من طريق النقل.

التأويل: صرف الكلام إلى معنىً محتمل موافق لما قبلها وما بعدها غير مخالف للكتاب والسنّة على طريق الاستنباط.

قيل: التفسير كشف ظاهر القرآن، والتأويل كشف باطنه.

وقيل: التفسير ما يتعلّق بالرواية، والتأويل ما يتعلّق بالدراية.

وقيل: التفسير ما يمكن إدراك البشر إلى حقائقه[1] ومعانيه، والتأويل ما يمتنع إدراكه ﴿وَما يَعْلَمُ[2] تَأويلَهُ إلّا اللهُ﴾.[3]

وقيل: التفسير ما لم يختلف فيه، والتأويل ما اختلف فيه.

وقيل: التفسير بيان الحقيقة، والتأويل بيان المقصود.

وقيل: التفسير ما لم يرد فيه من السلف إلّا وجه واحد، والتأويل ما ورد فيه وجوه كثيرة.[4]

وقيل: التفسير للمحكمات، والتأويل للمتشابهات.

السورة: قطعة من القرآن، معلومة الأوائل والأواخر.

وقيل: هي الطائفة من القرآن المترجمة التي أقلّها ثلاث آيات.

الآية: قطعة من السورة، محدودة الصدور والأعجاز.

المحكم: ما أحكمت عبارته بأن حفظت من الاحتمال.[5]

المتشابه: هو المشتبه[6] المحتمل للتأويل.

وقيل: المحكم ما لا يحتمل من التأويل إلّا وجهاً واحداً، والمتشابه ما احتمل أوجهاً.

وقيل: المحكم ما نعلم تأويله، والمتشابه ما استأثر الله بعلمه.

وقيل: المحكم ما لم تتكرّر[7] ألفاظه، والمتشابه ما تكرّرت.

وقيل: المحكم ما أمر الله به في كلّ كتاب، والمتشابه ما سواه.

وقيل: المحكم الناسخ، والمتشابه المنسوخ.

---

[1] حقائقه: صفاته، ب. [2] ما يعلم: لا يعلم، س. [3] آل عمران، 7. [4] كثيرة: كبيرة، ب. [5] الاحتمال: الإجمال، آ س. [6] المشتبه: المتشبتة، آ. [7] تتكرّر: يتكرّر، آ.

 | DOI:10.1163/9789004423367_009

المكّي: ما نزل من القرآن بمكّة.

المدني: ما نزل منه بالمدينة.

المجمل: اللفظ المتردّد بين معنيين فصاعداً من غير رجحان جهة على أخرى.

وقيل: ما لا يفهم معناه من لفظه على الإطلاق.

المفسّر: ما يفهم معناه من لفظه ولا يقبل التأويل.

وقيل: المفسّر والمحكم واحد.

المبيّن: الواضح بنفسه أو بغيره.

المأوّل: ما يبيّن[1] فيه احتمال الغلط.[2]

المطلق: اللفظ الدالّ على واحد غير معيّن.

المقيّد: ما يتناول ذاتاً غير موصوف بوصف.

العامّ: لفظ يستغرق جميع ما يصلح له بوضع واحد.

الخاصّ: ما يشمل بعض ما يتناوله اللفظ.

المنطوق: ما يفهم من اللفظ في محلّ النطق.

المفهوم: ما يفهم منه في غير محلّ النطق.

الناسخ: ما يبيّن انتهاء حكم سابق.

المنسوخ: ما انتهى حكمه بحكم آخر متأخّر عنه.

والله أعلم.

[1] يبيّن: بين، ب. [2] الغلط: اللفظ، ب.

الباب الثاني: في الحديث

علم الحديث: تتبّع أقوال رسول الله، صلّى الله عليه وسلّم، وأفعاله وما يتّصل بهما مما يفيد أحد الأحكام الخمسة.

الحديث: قول النبيّ، صلّى الله[1] عليه وسلّم، أو حكاية الصحابي عنه فعلاً مفيداً لحكم.

الصحيح: ما اتّصل سنده بنقل العدل الضابط عن مثله وسلم من[2] شذوذ وعلّة.

الحسن: ما لا يكون في إسناده متّهم.

وقيل: ما عرف مخرجه واشتهر رجاله.

وقيل: ما اشتهر راويه[3] بالصدق والأمانة، وقصر عن درجة رجال الصحيح حفظاً وإتقاناً بحيث لا يعدّ ما انفرد به منكراً.

وقيل:[4] ما لم يخل رجال إسناده عن مستور غير مغفّل في روايته.

الضعيف: ما لم يجتمع فيه شروط الصحيح والحسن.

المُسْنَد: ما اتّصل سنده من راويه[5] إلى منتهاه.

وقيل: ما نقل بالعَنْعَنَة من غير حذف واسطة.

المتّصل: ما اتّصل رواته، وكان كلّ واحد منهم قد سمعه ممن فوقه، سواء كان مرفوعاً إلى النبيّ، صلّى الله عليه وسلّم، أو موقوفاً[6] على غيره.

المرفوع: ما أضيف إلى النبيّ، صلّى الله عليه وسلّم[7]، من قول أو فعل أو تقرير[8]، سواء كان متّصلاً أو منقطعاً.

المُعَنْعَن: ما يقال في سنده: فلان عن فلان، لا بشرط البراءة عن التدليس.

الإفراد: ما تفرّد به عن جميع الرواة أو جمع[9] معيّنين أو شخص معيّن.

الآحاد: ما استقلّ بنقله الأفراد ولا يستوي أسفله أعلاه.

المُدرَج: ما أدرج في متنه أو إسناده شيء من[10] متن أو إسناد آخَرَين.

---

[1] الله: + تعالى، س. [2] من: عن، آ ب س. [3] راويه: رواته، آ ب س. [4] وقيل: + وقيل، آ. [5] راويه: رواتة، ب. [6] موقوفاً: مرفوعاً، آ. [7] صلّى [...] وسلّم: -، ب. [8] تقرير: تفسير، آ س. [9] جمع: جميع، آ س. [10] من: عن، آ س.

 | DOI:10.1163/9789004423367_010

المشهور: ما رواه جماعة من أهل العدالة والضبط وشاع عند أهل الحديث خاصّة دون غيرهم.

المستفيض: ما كان مشهوراً بين الناس، ولم يبلغ رتبة المتواتر.

الغريب: ما انفرد[1] به عدل ضابط.

العزيز: ما تفرّد به اثنان أو ثلاثة من أهل العدالة والضبط.

المصحَّف: ما وقع فيه الشبهة من جهة الكتابة.

المسلسل: ما تتابع فيه رجال الإسناد عند روايته على حالة أو صفة.

المختلف: أن يوجد حديثان متضادّان في المعنى ظاهراً فيجمع بينهما أو يرجّح أحدهما.

المُرْسَل: أن يقول التابعي: قال أو فعل رسول الله، صلّى الله عليه وسلّم، كذا، ولم يذكر الصحابي.

المعضل: ما سقط من سنده اثنان فصاعداً.

الشاذّ:[2] ما رواه الثقة مخالفاً لما رواه الناس.

وقيل: ما ليس له إلّا إسناد واحد شذّ به شيخ، ثقة كان أو غير ثقة.

المعلَّل: ما اطّلع فيه على ما يقدح في صحّته مع أنّ ظاهره السلامة منه.[3]

المدلّس: ما أخفي عيبه في إسناده لضعف فيه.

المضطرب: ما اختلفت الرواية فيه على وجوه مختلفة.

المقلوب: ما قلب إسناده للإشكال أو الترغيب.

المطعون: ما طعن في راويه إمّا فسقاً أو ذهولاً أو بدعة أو إسناداً.

الموضوع: كلّ متن مختلق على رسول الله، صلّى الله عليه وسلّم.

فصل

الإسناد: رفع الحديث إلى قائله.

السند: إخبار عن طريق المتن.

المتن: ما يتقوّم به معنى الحديث.

الاعتبار: التحرّي والاجتهاد في حال الإسناد ليطّلع[4] على المتابعات والشواهد.

[1] انفرد: تفرّد، آ س. [2] الشاذّ: الشان، آ. [3] منه: عنه، آ ب س. [4] ليطّلع: فيطلع، آ.

السماع: لقف الحديث عن لفظ الشيخ إملاءاً أو تحديثاً من حفظه أو كتابته.[1]

العَرْض: أخذ الحديث المقروء، هو أو غيره، على الشيخ كتابة[2] أو حفظاً.

الإجازة: إذن الشيخ لمعيّن أو غير معيّن في الرواية عنه.

المناولة: أن يُؤتِيَه أصل السماع أو الفرع المقابل به تمليكاً أو إعارة مقروناً بهما إذن الرواية عنه.

المكاتبة: كتابة مسموعه لغائب أو حاضر بخطّه[3] أو بإذنه مقرونة بالإجازة أو مجرّدة عنها.

الإعلام: إخبار[4] رواية كتاب أو سماعه مقتصراً عليه.

الوجادة: وجدان متن أو إسناد بخطّ شيخ معروف لا يصحّ إجازته منه.

الصحابي: كلّ مسلم رأى رسول الله، صلّى الله عليه وسلّم.

التابعي: كلّ مسلم عدلٍ لقي[5] صحابياً.

المولى:[6] المنتسب إلى شخص أو قبيلة بعتاقة أو حلف.[7]

التصحيح: كتابة "صح" فيما عرضَه الشكّ والخلاف ليدلّ على صحّته رواية ومعنىً.

التضبيب:[8] مَدّ خطٍّ كرأس الضاد المعجمة على ثابتٍ نقلاً فاسدٍ لفظاً أو معنىً، أو على ضعيف أو ناقص.

التمريض: مثله.

المؤتلف المختلف:[9] ما يتّفق في الخطّ دون اللفظ.

المتّفق المفترق: ما يتّفق في الخطّ واللفظ معاً.

الدارة: ما يجعل بين كلّ حديثين للتمييز.[10]

مثبِت السماع: من يكتب أسامي الطالبين يوماً فيوماً في جريدة.

والله أعلم.

---

[1] كتابته: كتابة، آ. [2] كتابة: كتاباً، آ. [3] بخطّه: لحفظه، آ، س. [4] إخبار: اختاره، س. [5] لقي: -، آ. [6] المولى: المدلى، آ، س. [7] حلف: خلف، آ. [8] التضبيب: التضبب، آ. [9] المختلف: -، آ. [10] للتمييز: للتميين، آ.

الباب الثالث: في الفقه

الفقه: ملكة استنباط ما لم يصرّح الشارع مما[1] صرّح به.

وقيل: تتبّع الأحكام الشرعية العملية المكتسبة[2] من أدلّتها التفصيلية.[3]

الطَهور: الطاهر في نفسه، المطهِّر لغيره.

المستعمل: ما أديّ[4] به عبادة أو انتقل إليه منع.

الجرية: ما يقابل طرفي النجاسة إلى حافتي النَهر.

التحرّي: بذل المجهود في طلب المقصود.

وقيل: تمييز الطاهر من النجس بما غلب على الظنّ.

الخبَث: عين[5] مستقذرة[6] تمنع[7] ملابستها رفع الحدث[8] والصلاة.

الاستطابة: طلب الراحة بخروج ما يخرج من السبيلين.

المَذيّ: ماء رقيق أصفر ذو لزوجة يخرج من لذّة النظر أو اللمس أو الفكر.

المَنيّ: ماء غليظ ذو تدفّق يلتذّ بخروجه.

الوَدْي: ماء كدر يخرج عقيب البول.

الدِباغ: نزع الفضلات عن الجلد بحِرِّيف.[9]

الحَدَث: أمر معنوي على سائر البدن أو أعضاء الوضوء مانع عن صحّة الصلاة.

الوضوء: طهارة مشتملة على أركان معدودة.

النيّة: الإرادة المتوجّهة نحو الفعل ابتغاء لمرضات الله وامتثالاً لأمره.

الاستنثار: استنشاق الماء ثمّ إخراجه بنفس الأنف.

التيمّم: ترتيب الوجه واليدين مرتّباً بقصد التطهير.

الحيض: دم يخرج من[10] رحم المرأة غير نفاس.

النفاس: دم يخرج عقيب الولادة.

---

[1] مما: بما، آ س. [2] المكتسبة: المكتسب، آ س. [3] التفصيلية: التفضيلية، آ. [4] أديّ: + أديّ، آ. [5] عين: غير، س. [6] مستقذرة: مستقذة، ب. [7] تمنع: يمنع، آ. [8] الحدث: الحديث، ب. [9] بحرّيف: بحرلف، آ س. [10] من: عن، آ س.

 | DOI:10.1163/9789004423367_011

الاستحاضة: دم تراه المرأة غير حيض ولا نفاس.

وقيل: دم يخرج من العاذل وهو عِرق في أدنى الرحم دون قعره.

الأذان: إعلام الناس وقت العبادة بأذكار مخصوصة في أزمان[1] معيّنة.

الترجيع:[2] إتيان الشهادتين سرّاً قبل إتيانهما جهراً.

التثويب: قول[3] "الصلاة خير من النوم" بعد الحَيْعَلَتَين.

الصلاة: عبادة ذات عقد وسلام.

وقيل: قرآن ضمّ إليها أفعال مخصوصة.

الركن: ما يتوقّف عليه وجود الشيء وتصوّره.

الشرط: ما يتوقّف عليه صحّة الشيء فقط.

الصحّة: استتباع الغاية.

البطلان: ضدّها.

الفساد: بمعناه.

الإجزاء:[4] هو الأداء الكافي لسقوط التعبّدية.

وقيل: سقوط القضاء.

الأداء: العبادة التي وقعت في وقتها المعيّن[5] ولم تسبق بأداء مختلّ.

القضاء: التي وقعت بعد وقتها ووجد فيه سبب وجوبها.

وقيل: إتيان العبادة بعد وقتها المقدّر استدراكاً لما فات.

الجماعة: ما اقتدى[6] فيها بإمام.

المسبوق: من لا يسع بين تكبيره وركوع إمامه قراءة فاتحة.

الجمعة: فرض عين لا يجب قضاءه.

العذر: ما يمنع وجوب القضاء.

الرخصة: ما شرع لعذر مع قيام المانع.

صلاة الخوف: صلاة بما يجوز فيها ما جرت العادة في الحرب.

صلاة الاستسقاء: صلاة يستسقى بها الغيث أو نبوع[7] عين انقطعت.

---

[1]أزمان: زمان، آ. [2]الترجيع: الرجيع، آ. [3]قول: قبل، آ. [4]الإجزاء: الإجراء، آ. [5]المعيّن: للعيّن، آ.
[6]اقتدى: أقدي، ب. [7]نبوع: ينوع، آ.

الجنائز: ما شرع للمحتضر والميّت وجوباً وندباً وإباحة.

الزكاة: إخراج جزء مقدّر من نصاب بنيّة شرعاً.

النصاب: ما يبلغ به المال حالة الكمال.

المعجّل: ما يقدّم[1] أداءه على وجوبه.

الخلطة: نصاب مشترك شائع أو متميّز[2] متّحد في أمور محصورة.

الفقير: من لا يقع ماله موقعاً من حاجته وكسب يليق به.

المسكين: من يقع ماله أو كسبه موقعها ولا يكفيه.

الغارم: المعسر المستدين لنفسه أو لله.

المؤلّفة: ضعفاء النيّة في الإسلام أو شرفاء يرجى بإعطائهم إسلام قومهم.

المكاتب: المملوك العاجز عن أداء ثمنه.

ابن السبيل: معسر منشئ سفر مباح أو مجتاز.

الصوم: إمساك يوم عما يوجب الفطر بنيّة.[3]

الاعتكاف: لبث[4] مسلم عاقل يحلّ[5] في المسجد زائداً[6] على الطمأنينة بالنيّة.

الحجّ: طواف الكعبة في وقت معيّن بأفعال مخصوصة.

وقيل: ما يتوقّف فرضه على استطاعة مالية[7] وأمن في السلوك إلى الكعبة وعرفات في أيّام معلومة.

العمرة: طوافها بأفعال مخصوصة غير مقيّدة[8] بوقت.

وقيل: ما يتوقّف فرضيّته[9] على الاستطاعة والأمن غير مقيّدة بوقت.

الإحرام: نيّة الحجّ أو[10] العمرة أو[11] هما في مكان معيّن.

الإفراد: أن يحجّ، ثم يعتمر.

التمتّع: أن يعتمر، ثم يحجّ.

القِران: أن يحرم بهما معاً.

الهَدى: ما يهدى من النعم إلى الحرم تقرّباً.

---

[1] يقدّم: يتقدّم، آ. [2] متميّز: متمين، آ. [3] بنية: بنيته، ب. [4] لبث: لبنت، آ. [5] يحلّ: محلّ، آ، س. [6] زائداً: -، آ س. [7] مالية: -، آ؛ بالنية، س. [8] مقيّدة: مفيدة، آ. [9] فرضيّته: فرضه، آ. [10] أو: و، آ. [11] أو: و، آ.

فصل

البيع: ما يوجب انتقال ملك مطلقاً بعوض.

وقيل: تمليك العين أو المنفعة على التأبيد بعوض مالي.

وقيل: معاوضة مال بمال ونحوه بالتراضي.

وقيل: مبادلة مال أو منفعة غير موقوف على التأبيد بعوض مالي بلا قرض وفسخ.

وقيل: نقل[1] عين مال معيّن أو في الذمة[2] من مالك إلى مالك بعوض بالاختيار حالاً أو مؤجّلاً.

المزابنة: بيع الشيء بالشيء لا مكيلاً ولا موزوناً.

المجازفة: كذلك.

المحاقلة: بيع الزرع بالحنطة.

المخابرة: المزارعة بالثلث أو[3] الربع أو نحو ذلك.

بيع الغرر[4]: بيع الخطر كالطائر في الهواء والسمك في الماء.

الملاقح: ما في بطون الأمّهات.

المضامين: ما[5] في أصلاب الآباء.

الخيار: تمكّن المشتري من إنفاذ البيع ونقضه.

النَجَش: الزيادة على شري الغير من غير احتياج.

الجنس الربوي: ما يكون فيه صلاحية الطعم مطلقاً أو جوهرية الأثمان.

القَبَض: تسليط المشتري على التصرّف في المبيع على حسب ماله.[6]

التَولية: جعل المشتري غيره[7] مكانه فيما اشترى به بقول شرعي.

الموقوف: ما وقف نفاذه على إذن المالك والمشتري.

الإقالة: نقض بيع مبرم.

العيب: ما ينقص[8] القيمة أو العين نقصاناً يفوت به غرض صحيح يكون الغالب في أمثال ذلك المبيع عدمه.[9]

[1] نقل: -، آ. [2] الذمة: الزمة، ب. [3] أو: و، آ. [4] الغرر: القسرو، آس. [5] ما: -، آس. [6] ماله: حاله، ب.
[7] غيره: غير، آ. [8] ينقص: ينقض، آ. [9] عدمه: عدم، س.

السَلَم: عقد على موصوف في الذمّة[1] ببدل[2] يعطى عاجلاً.

السَلَف: كذلك.

القرض: تمليك أهل التبرّع مالاً ليستردّ مثله متى شاء.[3]

القِراض: ما يدفع إلى غير ليتّجر فيه ويختصّ بجزء معلوم من الربح.

المضاربة: كذلك.

تفريق الصَفْقَة: أن يبيع رجل من رجل سلعتين، سمّى لكلّ واحدة[4] ثمناً معلوماً منفرداً.

الرهن: توثيق دين بعين.

التفليس: نداء الحاكم على أحد بالحجر أو بإفلاسه.[5]

الحجر: منع المالك عن تصرّف [في] ماله لحقّه[6] أو لحقّ غيره.

الصلح: رفع الخصومة بين المتداعيين.

صلح المفاوضة:[7] ما يجرى على غير العين[8] المدّعاة.

التحالف: أن يحلف كلّ من المتعاقدين أو الوارث.

الحوالة: إبدال دين بآخر.

الضمان: إلزام[9] حقّ على آخر.

الكفالة: إلزام[10] إحضار من يستحقّ حضوره مجلس الحكم بإذنه أو بإذن وليّه أو غير[11] يلزم موته ردّها.

الوكالة: نيابة فيما لا يتعيّن منه المباشرة بإيجاب مكلّف.

الشركة: خلط مالين لا يتميّز أحدهما عن الآخر بشرط.[12]

شركة العنان: أن يأذن كلّ واحد ممن له أهلية التوكيل والتوكّل للآخر[13] في التصرّف في مال مشترك.

الإقرار: الإخبار عن ثبوت حقّ سابق.

العارية: إباحة الانتفاع بعين يبقى معه بلا عوض.

وقيل: هبة المنفعة مع استبقاء ملك الرقبة.

الغصْب: الاستيلاء على حقّ الغير ظلماً.

---

[1] الذمة: الزمة، ب. [2] ببدل: ببذل، س. [3] شاء: اشاء، ب. [4] واحدة: واحد، آ س. [5] بإفلاسه: بفلانة، آ. [6] لحقّه: الحقّه، ب. [7] المفاوضة: المعارضة، آ؛ المعاوضة، س. [8] العين: المعين، آ. [9] إلزام: التزام، آ س. [10] إلزام: التزام، آ س. [11] غير: عين، ب. [12] بشرط: بشرطه، آ س. [13] للآخر: الآخر، آ س.

الشُفعة: حقّ تملّك[1] قهري بعوض يثبت للشريك القديم[2] على الحادث.

المساقاة: تسليم النخيل أو الكروم[3] للتعهّد بجزء من ثمرها.

المزارعة: تسليم الأرض إلى غير ليعمل عليها بشرط جزء لم يخرج.

الإجارة: تمليك المنفعة مدّة بعوض مالي.

الجعالة: التزام جعل على عمل لا بطريق الإجارة.

إحياء الموات:[4] تملّك أرض لا مالك لها بالعمارة.

الوقف: حبس[5] الأصل وتسبيل[6] المنفعة.

الهبة: تمليك منجّز تامّ بلا عوض.

اللقط: أخذ المال الضائع للحفظ أو التملّك.[7]

اللقيط: صبي منبوذ[8] لا كافل له.

الفرائض: هي قسمة المواريث.

الميراث:[9] انتقال الشيء إلى الإنسان بلا عقد ولا تبرّع.

المناسخة: أن يموت ورثة بعد ورثة، والميراث بعدلم يقسّم.

العَول: أن يزاد أجزاء الفريضة على الرؤوس.

العصبة: قرابة الرجل لأبيه، وبنوه وبنو أبيه.[10]

وقيل: من يأخذ كلّ التركة عند انفراده، والفاضل عن الفرض.

وقيل: كلّ ذكر ليس بينه وبين الميّت أنثى.[11]

الكلالة: أن يموت الرجل ولم يترك ولداً ولا ولدَ ولدٍ.

الوصية: تبرّع مضاف إلى ما بعد الموت.

العمري: أن يقول: جعلتُ لك هذا المال مدّة عمري أو عمرك؛ فإن متُّ قبلك فهو لك.

الرُقبي: كذلك.

الوصاية: استنابة[12] بعد الموت.

الإيداع: توكيل بحفظ المال.

---

[1] تملّک: تمليک، ب. [2] القديم: القويم، آ. [3] الكروم: الكرم، آ س. [4] الموات: الأموات، آ. [5] حبس: جنس، آ س. [6] تسبيل: سبيل، آ. [7] التملّک: التمليک، س. [8] منبوذ: منبوذة، ب. [9] الميراث: في ب كتبت في الهامش. [10] أبيه: انبه، آ. [11] أنثى: وأنثى، آ. [12] استنابة: استنابط، آ.

الفَيء: ما حصل من الكفّار بلا إيجاف خيل وركاب.

الغنيمة: ما يحصل منهم بإيجافها.

فصل

النكاح: تمليك بضع، وتملّك بإيجاب وقبول، وولّي وشاهدَي عدل.

الصداق: المال الواجب للمرأة على الرجل بالنكاح أو[1] الوطي.

الصدقة: ترادفه.

التفويض: إخلاء النكاح عن المهر بإذن من يستحقّه.

الخلع: فرقة بين الزوجين بعوض يأخذه الزوج.

التشطّر: رجوع نصف المهر إلى الزوج قهراً.

القسم: تقدير الرجل مدّة لبثه عند نسائه.

النشوز: بغض المرأة زوجها وخروجها عن طاعته.

الطلاق: رفع قيد النكاح بلا فسخ.

الإكراه: تهديد قاهر ذي عقوبة على محذور عظيم يقدر على إنقاذه[2] حالاً.

الرجعة: ردّ زوج لم يستوف عدد طلاقه، أو وليّ مجنون مطلّقة معيّنة[3] مجّاناً إلى نكاحه بلفظ منجّز.

الإيلاء: حلف[4] الزوج على ترك وطئ[5] زوجته الممكن في القبل، مطلقاً أو فوق أربعة أشهر في يمين.

الظهار: تشبيه مكلّف زوجته التي لم تصر بائنة أو جزئها بجزء محرم أنثى لم تحلّ له أبداً.

اللعان: يمين مؤكّدة بالشهادة واللعن أو[6] الغضب نفياً للولد أو دفعاً للتعزير[7] أو الحدّ.

القذف: سبّ[8] مكلّف مكلّفاً بصيغ[9] توجب الحدّ.

العدّة: مدّة تربّص الزوجة لبراءة الرحم أو التفجّع على الزوج.

الإحداد: ترك الزينة والطيب بعد خبر وفاة الزوج.

---

[1] أو: و، آ س. [2] إنقاذه: إنقاده، آ؛ إنفاذه، س. [3] معيّنة: معين، ب. [4] حلف: خلف، ب. [5] وطئ: الوطئ، آ س. [6] أو: و، آ. [7] للتعزير: للتغريس، س. [8] سبّ: سبب، آ. [9] بصيغ: بضيع، آ.

الاستبراء: مدّة تربّص الأَمَة عند حدوث مِلك اليمين أو زوال الفراش أو إرادة تزويجها إن وُطئت بشبهة أو حلّ.

الحضانة: القيام بمصالح من لا يستقلّ بأمره.

النفقة: مال يجب بمِلك أو زوجية أو قرابة.

فصل

الجراح:[1] جناية توجب[2] قَوَداً أو أرشاً.[3]

القتل: فعل في محلّ يتعقّبه زهوق روح المقتول به.

القصاص: حقّ إزهاق وجب لأولياء المقتول على القاتل في العمد المحض.

الدية: بدل[4] نفس الحرّ أو طرفه.

الغرّة: رقيق يبذل في قتل الجنين.

اللوث: أمارة تغلب على الظنّ صدق الوليّ في دعوى الدم.

القسامة: حلف من يستحقّ دم المقتول.

البغي: مخالفة فرقة ذي[5] شوكة يمكنهم المقاومة بمطاع الإمام العادل بتأويل باطل ظنّاً.

الرِدَّة: كفر المسلم المكلّف بقول أو فعل عناداً أو اعتقاداً أو استهزاءاً صريحاً.

الجزية: ما يقضى من المال على أهل الكتاب.

الزنا: إيلاج فرج في فرج محرّم قطعاً لعينه مشتهياً طبعاً بلا شبهة وكره وملك وظنّة وتحليل عالم.

المحُصَن: المكلّف الحرّ كلّه، المصيب في نكاح[6] صحيح قدر الحشفة بعدها.

السرقة: أخذ مال الغير خفية.

قاطع الطريق: مكلّف مسلم يعتمد[7] [على] القوّة في المغالبة[8] بالبعد عن الغوث.

المهادنة:[9] ترك الإمام أو نائبه قتال الكفّار لمصلحة.[10]

---

[1]الجراح: الحرام، آس. [2]توجب: تعجب، آ؛ تعحب، ب. [3]أرشاً: ارشادا، آ. [4]بدل: بذل، آ. [5]ذي: -، آ. [6]نكاح: ركاح، ب. [7]يعتمد: يعتمدة، آ. [8]المغالبة: المبالغة، آ. [9]المهادنة: المهاذنة، آب؛ المحاذنة، س. [10]لمصلحة: لمصحلة، ب.

المعاهدة والموادعة: توازيانها.

الأمان: إزالة الخوف عن الكافر من غير مال.

الذبيحة: حيوان مأكول لا يحلّ ميتة.

الحلال: ما لا ترجّح لأحد طرفيه على الآخر.

الجارح: كلّ محدّد[1] أنهر الدم.[2]

الأضحية: ما يتقرّب بذبحه إلى الله من النعم.

العقيقة: ما يضحّى للمولود من النعم المجزئة.

الجلّالة: التي تأكل العذرة.

المسابقة: مغالبة في نوع جنس تمريناً فيما هو الغرض[3] منه شرعاً.

اليمين: تحقيق ما لم يجب عقلاً بذكر اسم الله أو صفته.

النكول: الامتناع عن[4] اليمين.

الجرح: ردّ شهادة الشاهد بفسق أو مجادلة أو مسقط للمروءة.

التزكية: ضدّ الجرح.

القسمة: إفراز[5] حقّ كلّ ذي حقّ من الملك شرعاً.

النذر: التزام مكلّف مسلم قربة غير واجبة[6] على الأعيان بلفظ.

الكتابة: تعليق عتق بأداء مال منجّم.

النجوم: دفعات المال التي يؤدّيها.

التعجيز: أن يعجز المكاتب نفسه أو يعجز سيّده[7] فينقض المكاتبة.[8]

المستولدة:[9] أمة أتت بظاهر[10] تخطيط[11] علقت به من السيّد في ملكه.

التدبير: تعليق المكلّف[12] عتق ملكه بموته بصيغ[13] موضوعة له.

الإعتاق: إطلاق الرقيق من قيد الرقّ بعبارة مخصوصة.

والله أعلم.

---

[1] محدّد: محدود، آ. [2] أنهر الدم: انها لدم، ب. [3] الغرض: الفضل، ب. [4] عن: من، ب. [5] إفراز: افزار، آ.
[6] واجبة: واجب، آ. [7] سيّده: سنده، س. [8] المكاتبة: المكاتب، ب. [9] المستولدة: المستودلة، آ. [10] بظاهر: بظاهره، س. [11] تخطيط: تحطيط، آ؛ تحظيط، س. [12] المكلّف: المكاتب، آ. [13] بصيغ: بضبع، آ؛ ضيع، ب.

الباب الرابع: في أصول الفقه

أصول الفقه: معرفة دلائل الفقه إجمالاً، وكيفية الاستفادة[1] منها، وحال المستفيد.

الحكم: خطاب الله تعالى المتعلّق بأفعال المكلّفين بالاقتضاء أو[2] التخيير.

الخطاب: الكلام الذي يفهم[3] المستمع منه شيئاً.

التكليف: توجيه الخطاب إلى المخاطب.

وقيل: إلزام ما يشقّ.

الواجب: ما يُذمّ تاركه قصداً مطلقاً.

وقيل: ما ثبت وجوبه بدليل غير[4] قطعي.

وقيل: ما يجب عملاً، لا[5] اعتقاداً، ولا علماً.[6]

وقيل: ما[7] فرض بدليل فيه نوع شكّ.

وقيل: ما لا يسع تركه، ويلحق الإثم بتركه.

الفرض: يرادفه.

وقيل: ما يكفّر جاحده.

وقيل: ما ثبت وجوبه بدليل قطعي.

وقيل: ما يجب عملاً واعتقاداً[8] وعلماً.

الحتم والمركّب: كذلك.

النَدب: ما يُحمد فاعله ولا يُذمّ تاركه.

النافلة والسنّة: ترادفانه.

المباح: ما لا يتعلّق بفعله وتركه مدح ولا ذمّ.

وقيل: ما استوى فعله وتركه.

وقيل: ما اعتدل طرفاه.

---

[1] الاستفادة: الاستفاد، ب. [2] بالاقتضاء أو: بالاحتصار و، آ. [3] يفهم: يفهمه، ب. [4] غير: -، آ. [5] لا: لها، ب. [6] ولا علماً: وعلماً، آ. [7] ما: -، ب. [8] واعتقاداً: لا اعتقاداً، آ.

 | DOI:10.1163/9789004423367_012

الجائز: بمعناه.

المكروه: ما يمدح تاركه، ولا يذمّ فاعله.

الحرام: ما يذمّ فاعله شرعاً.[1]

المحظور: كذلك.

العزيمة: الحكم الثابت على وفق الدليل.

الرخصة: حكم ثبت على خلاف الدليل لعذر.

النَصّ: اللفظ المفيد المرتفع عن قبول التأويل.

وقيل: ما لا يحتمل إلّا تأويلاً واحداً.

وقيل: ما يستوي ظاهره وباطنه.

وقيل: ما تعرّى لفظه عن الشبهة ومعناه عن الشركة.

وقيل: ما رفع[2] في بيانه إلى أقصى غايته.

البيان: إخراج الشيء عن حيّز[3] الإشكال إلى فضاء الوضوح.

وقيل: هو الدليل الذي يتوصّل بصحيح النظر فيه إلى علم أو ظنّ.

وقيل: هو العلم الحاصل من الدليل.

الظاهر: ما احتمل معنيين، أحدهما أظهر من الآخر.

الأصل: هو المتّفق على تعليله.

وقيل: هو[4] الحكم إمّا بنصّ أو إجماع.

الفرع: ما لا يستقلّ بنفسه.

وقيل: المختلف في تعليله.

مفهوم المخالفة: ما يخالف حكم المسكوت عنه حكم المنطوق.

مفهوم الموافقة: ما يكون حكم مسكوته موافقاً لحكم منطوقه، أو هو أولى به، وقد يسمّى لحوى الخطاب.

[1] فاعله شرعاً: شرعاً فاعله، ب س. [2] رفع: وقع، ب. [3] حيّز: حيّرة، ب. [4] هو: -، ب.

فصل

العلم: تصديق جازم مطابق للواقع لموجب.

وقيل: صفة توجب لمحلّها تمييزاً[1] لا يحتمل النقيض بوجه.

وقيل: معرفة الشيء على ما هو به.

وقيل: هو انتفاء الخفاء.

اليقين: علم يحصل بعد استدلال ورويّة.[2]

الظنّ: تجويز أمرين أحدهما أظهر من الآخر.

الوهم: طرفه الآخر.

الشكّ: التردّد[3] بين أمرين لا مزيّة لأحدهما على الآخر.

السهو: ذهول[4] المعلوم إن خطر على البال، وتنبّه[5] صاحبه بأدنى تنبيه.

الجهل: اعتقاد الشيء على خلاف ما هو به.

الخطأ: الذهول من المعلوم محتاجاً إلى زيادة تنبيه.

الاشتقاق: ردّ لفظ إلى آخر لموافقته في حروفه الأصلية، ومناسبته في المعنى.

الترادف: توالي الألفاظ المفردة الدالّة على شيء واحد باعتبار واحد.

الاشتراك: تواطؤ[6] المعاني المختلفة تحت لفظ واحد باعتبار واحد.

الحقيقة: هي اللفظ المستعمل فيما وضع له في اصطلاح التخاطب.

المجاز: اللفظ المستعمل في معنىً غير موضوع له يناسب المصطلح.

الموافقة: اعتقاد حقّية الأمر.

المخالفة: اعتقاد فساده.

المختلفان: ما اختلفا في أخصّ صفاتهما.

الضدّان: ما[7] لا يسدّ أحدهما مسدّ الآخر.

وقيل: ما لا يتصوّر اجتماعهما في زمان واحد في مكان واحد[8] قوّة وفعلاً.

---

[1] تمييزاً: تميزاً، آ ب. [2] رويّة: رواية، آ. [3] التردّد: تردّد، ب. [4] ذهول: فعول، آ. [5] تنبّه: يتنبّه، آ ب س.
[6] تواطؤ: توالي، آ. [7] ما: -، آ. [8] واحد: آخر، آ.

الأمر: استدعاء الفعل بالقول ممن هو دونه.

وقيل: الطلب الجزم والاقتضاء[1] الحتم.[2]

وقيل: طلب اتّحاد المأخوذ.

النهي: القول الذي يستدعي ترك الفعل ممن هو دونه.

وقيل: هو[3] الدعاء إلى الإحجام.

النفي: القول الذي يدلّ على عدم المنفيّ.

الاستثناء: إخراج بعض الجملة بـ"إلّا" غير الصفة ونحوها.

الشرط: ما يتوقّف عليه تأثير المؤثّر.

النسخ: بيان انتهاء حكم شرعي متراخ.

فصل

السنّة: قول الرسول، صلّى الله[4] عليه وسلّم، أو فعله، أو تقريره.

المتواتر: كلّ خبر بلغت رواته[5] في الكثرة مبلغاً أحالت العادّة تواطؤهم على الكذب.

العدالة:[6] ملكة في النفس تمنع عن اقتران الكبائر والرذائل المباحة.

الخبر الصدق: ما عُلِم وجود مخبره بالضرورة أو الاستدلال.

الخبر الكذب: ما عُلِم خلافه ضرورة أو استدلالاً.

الاجتهاد: استفراغ الجهد في درك الأحكام الشرعية.

التقليد: قبول قول الغير بلا مطالبة حجّة.

الحجّة: ما دلّت على صحّة الدعوى.

الرأي: عبارة عن ظنّ قويّ.

المعرفة: إدراك صور الموجودات.

---

[1] الاقتضاء: الاقتصار، آ س. [2] الحتم: الختم، س. [3] هو: -، آ س. [4] الله: + تعالى، س. [5] رواته: رواية، آ.

[6] العدالة: + الة، آ.

البرهان: هو القياس المؤلّف اليقيني.

الإجماع: اتّفاق أهل الحلّ[1] والعقد من أمّة محمد، صلّى الله[2] عليه وسلّم.

القياس: حمل معلوم على معلوم في إثبات حكم لهما أو نفيه عنهما بناءاً على جامع.

وقيل: إثبات مثل حكم معلوم في معلوم آخر لاشتراكهما في علّة الحكم عند المثبت.

وقيل: ردّ الشيء إلى نظيره في الأصول الثلاثة.

العلّة: هي المعرّف للحكم.

الدلالة: الجمع بين الأصل والفرع بأمرٍ زائد على العلّة.

قياس العلّة: أن يُجمع[3] بين الأصل والفرع بعين العلّة.

قياس النسبة: أن يُجمع بين الأصل والفرع بما يوهم اشتماله[4] على العلّة من غير أن يلوح وجه المناسبة فيه.

تخريج[5] المناط: النظر في تعريف[6] علّية الحكم بالاستنباط.

تنقيح المناط: أن يُبيّن إلغاء الفارق.

الإخالة: ملائمة[7] الحكم والمصالحة[8] ومناسبة العلّة.

الحكم المناسب: ما يجلب للإنسان نفعاً، أو يدفع عنه ضرراً.

المنافي: ما يلزم منه مفسدة بتقدير ثبوت الحكم معه.

الملائم: ما أثّر جنس العلّة في جنس الحكم.

المؤثّر: ما ظهر تأثيره في الحكم.

الدوران: أن يحدث الحكم بحدوثه وينعدم بعدمه.

الطرد: وجود الحكم بوجود العلّة.[9]

وقيل: أن يثبت معه الحكم فيما عدا المتنازع فيه[10] فيثبت فيه إلحاقاً للفرد بالأعمّ الأغلب.

النقض: إبداء[11] الوصف بدون الحكم.

الكسر: عدم تأثير أحد الجزئين ونقض الآخر.

القلب: أن يربط خلاف قول المستدلّ على علّته إلحاقاً بأصله.

---

[1]الحلّ: الحق، س. [2]الله: + تعالى، س. [3]أن يجمع: أي يجمع، ب. [4]اشتماله: إشماله، ب. [5]تخريج: تخرج، آ. [6]تعريف: تعرّف، ب س. [7]ملائمة: ملاءكة، ب. [8]المصالحة: المصامحة، آ. [9]العلّة: الحكم، آ. [10]فيه: -، آ س. [11]إبداء: إيداء، آ س.

القول بالموجب: تسليم مقتضى قول المستدلّ مع بقاء الخلاف.

الفرق: جعل تعيّن[1] الأصل علّة أو الفرع مانعاً.

الترجيح: تقوية إحدى الأمارتين على الأخرى ليعمل بهما.

والله أعلم.

---

[1] تعيّن: تعيين، آس.

الباب الخامس: في أصول الكلام

علم الكلام: ما يبحث فيه عن ذات الله تعالى وصفاته وأحوال الممكنات في المبدأ والمعاد على قانون الإسلام.

وقيل: ما[1] يبحث فيه عن الأعراض الذاتية للموجود من حيث هو هو على قانون[2] الإسلام.

وقيل: علم يقتدر[3] معه على إثبات العقائد الدينية بإيراد الحجج ورفع الشبه.[4]

الواجب: ما اقتضت ذاته[5] وجوده في الخارج.

الممتنع: ما اقتضت ذاته عدم وجوده في الخارج.[6]

الممكن: ما لم تقتض[7] ذاته في الخارج وجوداً ولا عدماً.

الجوهر:[8] ممكن لم يكن في موضوع.

العرض: ممكن يكون في موضوع.

الوضع: كون الشيء مشاراً اليه إشارة حسّية.

الصورة النوعية: حالّة تكون مبدأ الآثار المختصّة بالنوع.

النفس النباتية: حالّة تكون مبدأ الاغتذاء[9] والنماء.

النفس الحيوانية: حالّة تكون مبدأً للحسّ[10] والحركة الإرادية.

النفس الطبيعية: بخلافها.

الطبيعة: بدء حركة وانتهاء سكون.

الطبع: هيئة يستكمل بها نوع من الأنواع[11]، كانت فعلية أو انفعالية.

الكمّ: ما يقبل الانقسام[12] لذاته.

الكيف: العرض الذي لا تكون ماهيّته بالقياس إلى الغير[13]، ولا يقتضي الانقسام لذاته.

المضاف: هو النسبة المتكرّرة.[14]

---

[1] ما: -، آ؛ علم، ب. [2] قانون: قان، آ. [3] يقتدر: يقدّر، ب. [4] الشبه: الشبهة، ب. [5] ذاته: + عدم، آ. [6] الممتنع ما اقتضت ذاته عدم وجوده في الخارج: -، آ. [7] تقتض: يقتض، آ س. [8] الجوهر: الجوهري، ب. [9] الاغتذاء: الاغتذال، آ؛ الاعتدال، س. [10] للحسّ: للجنس، ب. [11] من الأنواع: -، آ. [12] الانقسام: الاقسام، آ. [13] الغير: + ~~ولا يقضي~~، آ. [14] المتكرّرة: المكرّرة، آ ب.

 | DOI:10.1163/9789004423367_013

وقيل: ما ثبت[1] بثبوت الآخر.

الأين: هو الحصول في المكان.

متى: هو الحصول في الزمان.

الوضع: هيئة تحصل للجسم بسبب نسبة بعض أجزائه إلى البعض[2] وإلى الأمور الخارجة.

الملك: نسبة الشيء إلى ملاصق ينتقل بانتقاله.[3]

الفعل: هو التأثير.

الانفعال: هو التأثّر.

الحيّز: هو الفراغ المتوهّم المشغول بالشيء.

الشيء: ما يجوز أن يخبر عنه ويصحّ الدلالة عليه.

المعدوم: ما يصحّ أن يقال له "هل يوجد؟".

الجوهر الفرد: متحيّز لا يقبل القسمة.

الجسم: متحيّز قابل للقسمة.

العرض: هو الحالّ في المتحيّز.

الحركة: حصول الجوهر في حيّز قبل[4] حصوله في حيّز آخر.

السكون: حصوله بعد[5] حصوله في حيّز آخر.

الافتراق: حصول جوهرين في حيّزين بحيث[6] يمكن أن يتوسّطهما ثالث.

الاجتماع: حصولهما فيهما بحيث لا يمكن توسّطه.[7]

الوجود: هو الكون.

العدم: هو[8] انتفاء الكون.

الحال: صفة لموجود غير موصوف بالوجود والعدم.

الوجودي: ما ليس في نفس مفهومه وحقيقته نفي شيء.

العدمي: ما يكون كذلك.

الحدوث: كون الوجود مسبوقاً بالعدم.

---

[1] ثبت: يثبت، آ. [2] إلى البعض: -، ب. [3] بانتقاله: انتقاله، ب. [4] قبل: كذا في جميع النسخ، وهو غير مستقيم، والصحيح: بعد. [5] بعد: كذا في جميع النسخ، وهو لا يتّفق مع المعنى، والصحيح: قبل. [6] بحيث: الحيث، ب. [7] توسّطه: توسط، ب. [8] هو: -، ب.

وقيل: هو[1] الخروج من العدم إلى الوجود.

القدم: كون[2] الوجود غير مسبوق بالعدم.

الحدوث الإضافي: ما مضى من وجود[3] شيء أقلّ مما مضى من وجود آخر.

الحدوث المطلق: يقابله.

العلّة: ما يحتاج إليه وجود الشيء.

العلّة التامّة: جملة ما يحتاج إليه وجود الشيء.[4]

العلّة الغير التامّة: بعض ما يحتاج إليه وجود الشيء.

العلّة المادّية: ما يكون بها الشيء بالقوّة.

العلّة الصورية: ما يكون بها الشيء بالفعل.

العلّة الفاعلية: ما يفيد وجود الشيء.

العلّة الغائية: ما يكون لأجلها الشيء.

الفاعل المختار: هو الذي إن شاء فعل وإن شاء ترك.

فصل

العالم: كلّ موجود سوى الله تعالى.

الإيمان: تصديق الرسول بما علم مجيئه به ضرورة.

وقيل: إقرار باللسان، وتصديق بالجنان، وعمل بالأركان.

الاعتقاد: ربط القلب على الشيء بالتكليف.

وقيل: سكون النفس إلى الشيء.

التوحيد: العلم بأنّ الواحد واحد.

القضاء: علم الله تعالى بالكائنات من الأزل إلى الأبد على سبيل الإجمال.

القدر: علمه بها على سبيل التفصيل.

الإرادة: توافق العلم.

---

[1] هو: + ~~الوجود~~، ب. [2] كون: + ~~الشيء~~، آ. [3] وجود: جود، آ. [4] الشيء: + ~~الغير~~، آ.

وقالت المعتزلة: توافق الأمر.[1]

الإسلام: الانقياد لأحكام[2] الدين ظاهراً وباطناً.

الدين: اعتقاد الشيء على ما كان النحو عليه، حقّاً كان أو باطلاً.

وقيل: قول إلهي رادع للنفس يقوّمها[3] ويمنعها من الاسترسال فيما طبعت عليه.

وقيل: ما التزمه الإنسان لله عزّ وجلّ.[4]

الملّة: ما بيّنه[5] الله تعالى على لسان رسوله لعباده ذريعة[6] إلى جواره.[7]

الكفر: تغطية الحقّ بالباطل.

الشرع: هو الوضع الهادي إلى الحقّ.

المشروع: هو المأذون من قبل الشرع.

الرسول: هو المرسل من الله تعالى[8] إلى أمّته وإلى جميع الخلق.

وقيل: إنسان بعثه الله تعالى[9] إلى العباد لتبليغ[10] ما أوحي إليه.

الوحي: كلام إلهي منزل على نبيّ بواسطة الملك، أو بإلقاء معنىً في الروع.

الكلام: الألفاظ[11] التامّة المركّبة، وحروف دالّة على معانٍ مفيدة.

القرآن: هو الكلام المنزل للإعجاز بسورة منه.

المعجزة: فعل من أفعال الله تعالى[12] خارق للعادة مقترناً بدعوى النبوة موافقاً لدعواه[13] عند التحدّي مع عدم المعارضة.

العصمة: ملكة نفسانية تمنع عن الفجور.

الروحانيّون: أجسام لطيفة قادرة على التشكّل بأشكال مختلفة.

الكرامة: فعل خارج عن العادّة يظهر على شخص من غير دعواه، يرضاه[14] الله تعالى.[15]

العبادة: فعل خلّص لله[16] تعالى بالاختيار تعظيماً له بإذنه.

السنّة: ما واظب عليه رسول الله، صلّى الله[17] عليه وسلّم، مع الإشعار بجواز الترك.

الطاعة: امتثال أمر ثابت على حكم الواقعة.

---

[1] الأمر: -، آ. [2] لأحكام: لأحكم، آ. [3] يقوّمها: نقومها، آ. [4] وقيل قول إلهي رادع [...] لله عزّ وجلّ: -، ب. [5] بيّنه: يبيّنه، آ س. [6] ذريعة: وزيعة، آ. [7] جواره: وجواده، س. [8] تعالى: -، ب. [9] تعالى: -، ب. [10] لتبليغ: لتبلغ، آ. [11] الألفاظ: ألفاظ، آ. [12] تعالى: -، ب. [13] لدعواه: لداعواه، آ. [14] يرضاه: يرضى، ب. [15] تعالى: -، ب. [16] لله: الله، آ. [17] الله: + تعالى، س.

المعصية: مخالفة أمر الله تعالى مع التزام المؤاخذة.

الرحمة: إرادة الخير من الله لأهلها.

الغضب: إرادة العقوبة لأهلها.

التوفيق: خلق القدرة على الطاعة.

الخذلان: خلق القدرة على المعصية.

النعمة: ما كانت عاقبتها محمودة.

النقمة: ما كانت عاقبتها مذمومة.

وقيل: ما يوجد منها العذاب.

الإمامة:[1] هي الرئاسة العامّة في الدين والدنيا. والله أعلم.

[1] الأمامة: الأمانة، آ.

الباب السادس: في الجدل

علم الجدل: صناعة نظرية يستفاد منها كيفية المناظرة وشرائطها صيانة عن الخبط في البحث، وإلزاماً للخصم وإفحامه.

وقيل: قانون يفيد عرفان القدر الكافي من الهيئات وأقسام الاعتراضات والجوابات، الموجّهات منها[1] وغير الموجّهات.

المناظرة: النظر[2] بالبصيرة من الجانبين في النسبة بين الشيئين إظهاراً[3] للصواب.

المجادلة: تعارض يجري بين متنازعين[4] فصاعداً إمّا لتحقيق[5] حقّ أو تغليب ظنّ أو إبطال باطل.

الأدب: محافظة ما يجب محافظته في الشيء لأجل المباشرة والشروع فيه.

البحث: إثبات النسبة الإيجابية أو السلبية بين الشيئين بطريق الاستدلال.

الإشارة: دلالة[6] عقلية أو حسّية[7] إلى الشيء بحيث لا يشاركه فيها غيره.

الضلالة: سلوك طريق لا يوصل إلى المطلوب.

وقيل: فقدان ما يوصل إلى المطلوب.

الحيرة: ما ينشأ عن تعارض الأدلّة.

الفهم: تصوّر المعنى من لفظ[8] المخاطب.

التفهيم: إيصال المعنى إلى ذهن السامع بواسطة اللفظ.

الإلهام: إلقاء معنىً في الروع بطريق الفيض.

الصواب: ما يكون مطابقاً للأمر في نفسه.

الفكر: حركات تخييلية في الذهن.

وقيل: انتقال النفس في المعاني انتقالاً بالقصد لطلب علم أو ظنّ.

---

[1] منها: لمنها، آ. [2] النظر: -، آ. [3] إظهاراً: إظهار، ب. [4] متنازعين: المتنازعين، آ س. [5] لتحقيق: التحقيق، آ.
[6] عقلية: + عقليلة، آ. [7] حسّية: حسبيّة، ب. [8] لفظ: اللفظ، آ.

 | DOI:10.1163/9789004423367_014

فصل

الدليل: ما يؤدّي إلى إدراك المطلوب.

وقيل: ما يلزم من[1] العلم به العلم بشيء آخر.

وقيل: ما يمكن التوصّل بصحيح النظر فيه[2] إلى مطلوب خبري.

وقيل: هو كلّ أمر صحّ أن يتوصّل بصحيح النظر فيه[3] إلى علم مالم يُعلم باضطرار.[4]

الدالّ: هو المخبر عن الدلالة المنصوبة.

المدلول: هو المطلوب عن الدليل.

المدلول له:[5] من يُنصب لأجله[6] الدليل.

المستدَلّ: الناظر الطالب للعلم.

وقيل: هو السائل نصب الدليل.

المستدِلّ: هو المطالِب بالدلالة.

المستدَلّ له: هو الحكم المطلوب عن النظر.

المستدَلّ عليه: هو الحكم المدلول بالدليل.

الاستدلال: هو البحث والنظر.

وقيل: مسألة السائل عن الدليل.[7]

الأمارة: ما يلزم من العلم به الظنّ بوجود المدلول.

أركان الشيء: أجزاؤه في الوجود التي لا تحصل إلّا بحصولها داخلة في حقيقته محقّقةً لهويته.

العلّة: ما يتوقّف عليه الشيء.[8]

وقيل: ما يقتضي الحكم، إمّا تامّةً أو ناقصةً.

العلّة التامّة: جملة ما يتوقّف عليه وجود الشيء.

العلّة الناقصة: بعض ما يتوقّف عليه وجود الشيء.

الشرط: ما يوجد الحكم بوجوده.

---

[1] من: إلى، آس. [2] النظر فيه: الظرفية، ب. [3] النظر فيه: الظرفية، ب. [4] باضطرار: + فيه، آس. [5] المدلول له: المدلو له، ب. [6] لأجله: لأجل، آ. [7] الاستدلال [...] عن الدليل: -، آ. [8] الشيء: شيء، آ.

التعليل: تبيين علّية[1] الشيء الذي يطلب إثباته أو نفيه لينتقل الذهن من العلم بها إلى العلم بالمعلوم، ويقال له: الاستدلال.

وقيل: الاستدلال[2] هو انتقال الذهن من الأثر إلى المؤثّر، والتعليل انتقال الذهن من المؤثّر إلى الأثر.[3]

وقيل: الاستدلال تقرير ثبوت الأثر لإثبات المؤثّر، والتعليل تقرير ثبوت المؤثّر لإثبات الأثر.

الملازمة: هي كون الحكم مقتضياً لآخر.

اللزوم[4] **والتلازم**: كذلك.

الدوران: ترتّب الشيء على الشيء الذي له صلوح[5] العلّية وجوداً أو عدماً أو معاً.[6]

الدائر: هو المرتّب.

المدار: هو المرتّب عليه.

المناقضة: منع مقدّمة معيّنة من مقدّمات دليل الخصم.

المقدّمة: ما توقّف عليه صحّة الدليل.

المعارضة: إقامة الدليل على خلاف ما أقام الدليل عليه الخصم.

النقض: تخلّف الحكم عن الدليل.

السند:[7] ما يكون المنع مبنياً عليه.

معارضة المثل: أن يكون[8] صورة دليل السائل مثل صورة دليل المعلّل.

معارضة القلب: أن يكون دليل السائل عين دليل المعلّل.[9]

وقيل: مشاركة الخصم المستدلّ في علّة الحكم.

معارضة الغير: أن لا يكون دليل السائل عين دليل المعلّل، ولا صورته صورتَه.

المناسب الغريب: أن يرى[10] مقروناً بمعنىً مناسب من غير إجماع أو دلالة نصّ على ثبوته به أو[11] تأثير جنسه في عينه، أو تأثير جنسه في جنس ذلك.

الملائم: ما ظهر تأثير جنسه في جنس الحكم.

---

[1] علّية: علّة، آ س. [2] الاستدلال: -، آ س. [3] والتعليل [...] إلى الأثر: -، آ. [4] اللزوم: الملزوم، س. [5] صلوح: صلاح، س. [6] معاً: + ~~الدائر هو المرتب~~، ب. [7] السند: كذا في جميع النسخ، وفي آداب البحث للسمرقندي: المستند. [8] أن يكون: + أن يكون، س. [9] معارضة القلب [...] دليل المعلّل: كتبت في هامش ب. [10] أن يرى: أن ترى، ب. [11] أو: و، آ.

المانع: ما يلزم من وجوده انتفاء الحكم.

مانع السبب: ما يُخلّ بحكمة السبب.

مانع الحكم: ما يشتمل على حكمة هي بمعزل عن حكمة[1] السبب مقتضاها نقيض[2] مقتضى السبب.

والله أعلم.

[1] حكمة: الحكم، آ. [2] نقيض: يقتضى، ب.

الباب السابع: في النحو

علم النحو: معرفة أحوال الكلم وكيفية تركيباتها.

وقيل: هو العلم المستنبط بالاستقراء أو القياس من كتاب الله تعالى والكلام الفصيح.

الكلمة: هي اللفظة[1] الدالّة على معنىً[2] مفرد بالوضع.

وقيل: لفظ وضع لمعنىً مفرد.[3]

وقيل: هي اللفظ الموضوع للمعنى مفرداً.

اللفظ: ما يتلفّظ به الإنسان أو في حكمه، مهملاً كان أو مستعملاً.

الوضع: تخصيص شيء[4] بآخر[5] ليفهم الثاني عند إطلاق الأوّل أو تخيّله.

المعنى المفرد: ما لا يراد بجزء لفظه حالة الجزئية جزء معناه.

الكلام: هو المركّب من كلمتين أسندت إحداهما إلى[6] الأخرى.

وقيل: هو ما تضمّن كلمتين بالإسناد.[7]

والإسناد: نسبة أحد الجزئين إلى الآخر ليفيد المخاطب فائدة يصحّ السكوت عليها.

الاسم: ما دلّ على معنىً في نفسه دلالة مجرّدة[8] عن الاقتران بزمان.

وقيل: عن الاقتران بأحد الأزمنة الثلاثة.

المعرب: المركّب الذي لم يشبه مبنيّ الأصل مختلفاً آخره باختلاف العوامل لفظاً أو تقديراً.

الإعراب: ما اختلف آخره به ليدلّ على المعاني المعتورة عليه.

العامل: ما به يتقوّم المعنى المقتضي للإعراب.

غير المنصرف:[9] ما يمتنع صرفه عند عدم الضرورة لعلّة.

العدل: خروج الاسم عن صيغته الأصلية تحقيقاً[10] أو تقديراً.[11]

---

[1] اللفظة: اللفظ، آ. [2] معنى: + ما، آ. [3] مفرد: لمفرد، آ. [4] شيء: + بشيء، آ؛ + ~~بشيء~~، س. [5] بآخر: آخر، آ. [6] إلى: على، آ. [7] هو المركّب [...] ما تضمّن كلمتين بالإسناد: هو المركّب من كلمتين بالإسناد، س. [8] مجرّدة: + ~~على~~، ب. [9] المنصرف: المتصرّف، آ. [10] تحقيقاً: + كثلاث و مثلّث، س. [11] تقديراً: + كعمر و باب قطام، س.

 | DOI:10.1163/9789004423367_015

فصل

المرفوعات: ما اشتمل على علم الفاعلية.

الفاعل: ما كان المسند إليه من فعل[1] أو شبهه مقدّماً عليه أبداً.

وقيل: ما أسند إليه الفعل أو شبهه وقدّم عليه على جهة قيامه به.

مفعول ما لم يسمّ فاعله: كلّ مفعول حذف فاعله وأقيم هو مقامه.

المبتدأ: الاسم المجرّد عن ملابسة العوامل اللفظية معنًى من حيث هو اسم للإسناد إليه، والصفة المعتمدة على أحد حرفي الاستفهام والنفي رافعة لظاهر أو ما يجري مجراه.

الخبر: هو المجرّد المسند إلى ما تقدّمه لفظاً أو تقديراً.

خبر إنّ وأخواتها:[2] هو المسند بعد دخول هذه الحروف.

خبر لا لنفي الجنس: هو المسند بعد دخولها.

اسم ما ولا المشبّهتين بليس: هو المسند إليه بعد دخولهما.

فصل

المنصوبات: ما اشتمل على علم المفعولية.

المفعول المطلق: هو اسم ما فعله فاعل فعل مذكور بمعناه.

المفعول به: هو ما وقع عليه فعل الفاعل.

المنادى: هو المطلوب إقباله بحرف نائب مناب "أدعو" لفظاً أو تقديراً.

الترخيم: حذف في آخر المنادى للتخفيف.

المندوب: هو المُتَفجّع عليه بياء أو واو.[3]

المفعول فيه: ما فعل فيه فعل مذكور من زمان أو مكان.

الظرف: بمعناه.

المفعول له: ما فعل لأجله فعل مذكور.

---

[1] فعل: -، ب. [2] أخواتها: خواتها، ب. [3] واو: وا، آ، ب، س.

المفعول معه: مذكور[1] بعد الواو لمصاحبة معمول فعل لفظاً أو معنىً.

الحال: ما يبيّن هيئة الفاعل أو المفعول به لفظاً أو معنىً.

التمييز: ما يرفع الإبهام المستقرّ عن ذات مذكورة أو مُقدّرة.

الاستثناء: إخراج بعض ما يتناوله اللفظ.

المستثنى المتّصل: هو المخرج من متعدّد لفظاً أو معنىً.[2]

المستثنى[3] المنقطع:[4] هو المذكور بعد حرف الاستثناء غير مخرج.

خبر كان وأخواتها: هو المسند بعد دخولها.

المنصوب بلا النافية: هو المسند إليه بعد دخولها يليها نكرة مضافاً أو مشبّهاً به.

خبر ما ولا المشبّهتين بليس: هو المسند به بعد دخولهما.

فصل

المجرورات: ما اشتمل على علم المضاف إليه.

المضاف إليه: كل اسم نسب إليه شيء بواسطة حرف الجرّ لفظاً أو تقديراً مراداً.

الإضافة المعنوية: هي أن يكون[5] المضاف غير صفة مضافة إلى معمولها.

الإضافة اللفظية: أن يكون[6] صفة مضافة إلى معمولها.

التوابع: كلّ ثانٍ بإعراب سابقه من جهة واحدة.

النعت: تابع يدلّ على معنىً في متبوعه مطلقاً لتخصيص أو توضيح.

العطف: تابع مقصود بالنسبة[7] مع متبوعه يتوسّط بينهما حرف عطف.

التأكيد: تابع[8] يقرّر أمر المتبوع في النسبة أو الشمول.

التأكيد اللفظي: تكرير اللفظ بعينه لتقرير النسبة.

التأكيد المعنوي: تكرير اللفظ بألفاظ مخصوصة موضوعة[9] لذلك.[10]

---

[1]مذكور: مذكو، ب. [2]المستثنى المتّصل [...] أو معنىً: -، ب. [3]المستثنى: + ~~المنقع~~، آ. [4]المنقطع: المنقع، ب. [5]يكون: تكون، آ. [6]يكون: تكون، آ س. [7]بالنسبة: بما نسب، آ. [8]تابع: ~~يدلّ~~، آ. [9]موضوعة: -، ب. [10]لذلك: كذلك، آ.

البدل: تابع مقصود بما نسب إلى المتبوع دونه.

بدل الكلّ: ما يكون مدلوله مدلول الأوّل.

بدل البعض: ما يكون مدلوله بعض مدلول الأوّل.

بدل الاشتمال: ما يكون بينه وبين المبدل منه ملابسة غير البعضية و[1] الكلّية.

بدل الغلط: ما تقصد إليه بعد أن غلطت بغيره.

فصل

المبنيّ: ما ناسب مبنيّ الأصل أو وقع غير مركّب.

المضمر: ما وضع لمتكلّم أو مخاطب أو غائب تقدّم ذكره لفظاً أو معنىً أو حكماً.

المنفصل: هو[2] المضمر المستقلّ بنفسه.

المتّصل: ما لم يستقلّ بنفسه.

اسم الإشارة: ما وضع لمشار إليه.

الموصول: ما لا يتمّ جزءاً إلّا بصلة وعائد.

أسماء الأفعال: ما كان بمعنى الأمر أو الماضي.

أسماء الأصوات: كلّ لفظ حُكي به صوت أو صُوّت به للبهائم.

المركّبات: كلّ اسم من كلمتين ليس بينهما نسبة أصلاً.

الكنايات: كلّ لفظ كني به عن العدد أو الحديث أو الفعل.

المعرفة:[3] ما وضع لشيء بعينه غير متناول غيره بوضع[4] واحد.

النكرة: ما وضع لشيء لا بعينه.[5]

أسماء العدد: ما وضع لكمّية آحاد الأشياء.

المؤنّث: ما فيه علامة التأنيث لفظاً أو تقديراً.

المذكّر: بخلافه.

---

[1] و: أو، س. [2] هو: -، آ س. [3] المعرفة: كذا في جميع النسخ. والصحيح: العَلَم. [4] بوضع: بواضع، آ. [5] لشيء لا بعينه: لا بشيء بعينه، آ.

التأنيث الحقيقي: ما بإزائه ذكر من الحيوان.

اللفظي: بخلافه.

المثنّى: ما لحق آخره علامة التثنية ليدلّ على أنّ معه مثله من جنسه.

المجموع: ما دلّ على آحاد مقصودة بحروف مفرده بتغيير[1] ما.

جمع التكسير: ما تغيّر[2] فيه بناء واحده.

جمع الصحيح: بخلافه.

جمع القلّة: ما يطلق على العشرة فما دونها بطريق الحقيقة.[3]

جمع الكثرة: ما يطلق على ما فوق العشرة حقيقة.

المصدر: اسم الحدث الجاري على الفعل لبيان مدلوله.

اسم الفاعل: ما اشتقّ من فعل لمن قام به بمعنى الحدوث.

اسم المفعول: ما اشتقّ من فعل لمن وقع عليه ذلك الفعل.

الصفة المشبّهة: ما اشتقّ من فعل لازم لمن قام به على معنى الثبوت.

اسم التفضيل: ما اشتقّ من فعل لموصوف [بزيادة على غيره].

فصل

الفعل: ما دلّ على معنًى في نفسه مقترناً[4] بأحد الأزمنة الثلاثة.

الأمر: صيغة يطلب بها الفعل من الفاعل المخاطب بحذف حرف المضارعة.

النهي: صيغة يطلب بها ترك الفعل منه بدخول حرفه عليه.

فعل ما لم يسمّ فاعله: هو ما[5] حذف فاعله فيه.

أفعال القلوب: ما وضعت للشكّ أو[6] اليقين.

الأفعال الناقصة: ما وضعت لتقرير الفاعل على صفة.

أفعال المقاربة: ما وضعت[7] لدنوّ الخبر رجاءً أو حصولاً أو أخذاً فيه.

[1] بتغيير: بتغير، آ. [2] تغيّر: نفد، آ؛ نفد، ب س. [3] الحقيقة: الحقيقية، ب. [4] مقترناً: مقترن، آ س. [5] ما: -، ب. [6] أو: و، آ. [7] وضعت: وضع، آ س.

أفعال[1] التعجّب: ما وضعت لإنشاء التعجّب.
فعلا المدح والذمّ: ما وضع[2] لإنشاء مدح أو ذمّ.

فصل

الحرف: ما نتوقّف[3] دلالتها على معناه الإفرادي على متعلّق باعتبار الوضع.
حروف الجرّ: ما وضع للإفضاء بفعل أو شبهه أو معناه إلى ما يليه.
الحروف المشبّهة: حروف تدخل على الجملة الاسمية، فتنصب اسمها وترفع خبرها.
الحروف العاطفة: ما يعطف بها مفرد أو مركّب[4] على آخر.
حروف التنبيه: ما ينبّه بها على تحقيق الأمر.
حروف النداء: ما ينادى بها.
حروف الزيادة:[5] ما تقع زوائد لتصحيح وزن أو سجع[6] أو تأكيد.
حروف الصلة: ترادفها.

[1] أفعال: + المقاربة، ب. [2] وضع: وضعا، س. [3] نتوقّف: يتوقّف، ب س. [4] مفرد أو مركّب: مفرداً أو مركّباً، آ؛ مفرداً و مركّب، س. [5] الزيادة: الزوائد، آ. [6] سجع: شجع، آ.

الباب الثامن: في الصرف

علم الصرف: أصول يتعرّف بها أحوال أبنية الكلم التي ليست بإعراب.

الإلحاق: زيادة حرف في الكلمة لتصير على هيئة أصلية لكلمة فوقها في عدد الحروف الأصول ليعامل معاملته.[1]

الصحيح: ما ليس فيه حرف علّة.

المعتلّ: ما فيه حرف علّة.

المثال: ما اعتلّ فاؤه.[2]

الأجوف: ما اعتلّ عينه.

ذو الثلاثة: مثله.

الناقص: ما اعتلّ لامه.

ذو الأربعة: كذلك.

اللفيف: ما فيه حرفا علّة.

اللفيف المقرون: ما اعتلّ فاؤه وعينه[3]، أو عينه ولامه.

اللفيف المفروق: ما اعتلّ فاؤه ولامه.

الثلاثيّ:[4] ما كان على ثلاثة أحرف.

الرباعيّ: ما كان ماضيه[5] على أربعة أحرف.

والخماسيّ: ما كان ماضيه[6] على خمسة أحرف.

أفعال الطبائع: ما جبل عليها الإنسان أو صارت له ملكة بالتكرار.[7]

الفعل اللازم: ما لزم الفاعل ولم يجاوزه.

المتعدّي: ما يتوقّف[8] فهمه على متعلّق.

الواقع والمجاوز:[9] بمعناه.

---

[1] معاملته: معاملة، آ. [2] فاؤه: فاه، آ. [3] عينه: + ~~أو لامه~~، س. [4] الثلاثي: الثاني، آ. [5] ماضيه: -، ب س.
[6] ماضيه: -، ب س. [7] بالتكرار: باتكرار، آ. [8] يتوقّف: + في، ب. [9] المجاوز: + ~~مثبتاً~~، س.

 | DOI:10.1163/9789004423367_016

فصل

الآلة: كلّ اسم اشتقّ من فعل اسماً لما يستعان به في ذلك الفعل.

المصغّر: المزيد فيه ليدلّ[1] على تقليل.

المنسوب: الملحق بآخره ياء مشدّدة ليدلّ على نسبته إلى المجرّد عنها.

الوقف: قطع الكلمة عما بعدها.

الروم: أن تأتي بالحركة خفيّة.[2]

الإشمام: أن تضمّ الشفتين بعد الإسكان.

المقصور: ما آخره ألف مفردة.[3]

الممدود: ما يكون آخره ألف بعدها همزة.

ذو الزيادة: كلمة زيد[4] فيها حرف فصاعداً من حروف الزيادة.

حروف الزيادة: ما يزاد لغير الإلحاق والتضعيف.

الإمالة: أن تنحى[5] بالفتحة نحو الكسرة لقصد المناسبة.

الإعلال: تغيير حروف[6] العلّة للتخفيف.

الإدغام: أن تأتي بحرفين ساكن فمتحرّك من مخرج واحد بلا فصل.

الحروف المجهورة: ما ينحصر جري النفس مع تحرّكه.

المهموسة: بخلافها.

الشديدة: ما ينحصر جري صوته عند إسكانه في مخرجه.

الرخوة:[7] بخلافها.

ما بين الرخوة والشديدة: التي لا يتمّ لها الانحصار ولا الجري.

المطبقة: ما ينطبق على مخرجه الحنك.

المنفتحة: بخلافها.

المستعلية: ما يرتفع اللسان بها إلى الحنك.

المنخفضة: بخلافها.[8]

---

[1] ليدلّ: ليدخل، آ. [2] خفية: خفيفة، آ. [3] مفردة: مفرد، آ. [4] زيد: زيدت، آ ب س. [5] تنحى: تنحو، آ س.
[6] تغيير حروف: تغيّر حرف، ب. [7] الرخوة: الرجوة، آ. [8] المستعلية [...] بخلافها: -، ب.

حروف الذلاقة: ما لا ينفكّ رباعي ولا خماسيّ عن شيء منها.

المصمتة: بخلافها.

حروف القلقلة: ما ينضمّ إلى الشدّة فيها ضغط في الوقت.

حروف الصفير:[1] ما يصفر بها عند تلفّظها.

والله أعلم.

[1] الصفير: الصفر، آ.

الباب التاسع: في المعاني والبيان

علم المعاني: تتبّع خواصّ تراكيب الكلام في الإفادة وما يتّصل بها من الاستحسان وغيره، ليحترز[1] بالوقوف عليها عن الخطأ في تطبيق الكلام على ما يقتضي الحال ذكره.

خاصّية التركيب: ما يفيده التركيب[2] لا بمجرّد الوضع.

التطبيق: إصابة المخبر في بيان الحجّة.

الحال: الأمور الداعية إلى التكلّم على الوجه[3] المخصوص.

مقتضى الحال: ما يقتضيه الأمور الداعية إلى ذلك.

الصناعة: ملكة نفسانية يقتدر بها الإنسان على استعمال موضوعات ما نحو[4] غرض من الأغراض على سبيل الإرادة صادرة عن بصيرة بحسب التمكّن منها.

البلاغة في الكلام: مطابقته[5] لمقتضى الحال مع فصاحته.

بلاغة المتكلّم: بلوغه في تأدية المعنى حدّاً له[6] اختصاص بتوفية خواصّ التراكيب حقّها، وإيراد أنواع التشبيه والمجاز والكناية على وجهها.

وقيل: بلوغه في كلامه لعبارة كنه مراده مع إيجاز بلا إخلال، وإطالة بلا إملال.

وقيل: ملكة يقتدر بها على تأليف كلام بليغ.

الفصاحة: صوغ الكلام على وجه له توفية بتمام الإفهام لمعناه وتبيين[7] المراد وتزيين الألفاظ بما يقرب فهمه، ويعزب نظمه، ويعذب استماعه، ويعجب ابتداعه، ويدلّ مطالعه على مقاطعه[8]، وينمّ مبادئه على تواليه لا باستعمال الشوارد التي لا تفهم والأوابد التي لا تعلم.

فصاحة المتكلّم: ملكة نفسانية يقتدر بها على التعبير عن المقصود بلفظ فصيح.

الفصاحة اللفظية: خلوص الكلام عن ضعف التأليف، وتنافر الكلمات والتعقيد.

الفصاحة المعنوية: خلوصه عن التعقيد.

التعقيد: أن لا يكون الكلام ظاهر[9] الدلالة على المراد لخلل.

التنافر: إيراد الكلام على وجه لا يسهل على اللسان، إمّا لبعد شديد بين المخرجين، أو لقرب بليغ بينهما.

---

[1] ليحترز: + عن، س. [2] ما يفيده التركيب: -، آ. [3] الوجه: وجه، ب. [4] نحو: -، آ س. [5] مطابقته: مطابقة، آ. [6] له: -، آ. [7] وتبيين: ويتبيّن، ب. [8] مقاطعه: مطاطعه، آ. [9] ظاهر: ظاهرة، آ.

 | DOI:10.1163/9789004423367_017

فصل

الخبر: الكلام المحتمل للصدق والكذب.

وقيل: المحتمل للتصديق والتكذيب.

وقيل: الكلام المفيد بنفسه إضافة أمر من الأمور إلى أمر من الأمور[1] نفياً وإثباتاً.

وقيل: الكلام المقتضي بصريحه نسبة معلوم إلى معلوم بالنفي والإثبات.

لازم الخبر: احتمال الصدق والكذب.

صدق الخبر: مطابقته[2] للواقع.

وقيل: لاعتقاد[3] المخبر.

وقيل: لظنّه.

كذب الخبر: عدم مطابقته له.[4]

الكلام الابتدائي: ما خوطب به خالي الذهن.

وقيل: ما يحكم فيه بالمسند للمسند إليه حكماً خالياً عن المؤكّدات.

الكلام الظنّي: ما نفي به شكّ العالم.

وقيل: ما يحكم فيه حكماً مؤكّداً بنوع تأكيد لظنّ طلب في المخاطب.[5]

الكلام الإنكاري: ما ردّ به حكم المخالف.

وقيل: ما يكون الحكم فيه مؤكّداً لردّ المخاطب إلى حكم المخاطب.

الخطاب العامّ: ما يخاطب به غير معيّن.

الالتفات: نقل الكلام من الغيبة والخطاب والتكلّم، كلّ منها إلى الآخر أو الآخرين.

وقيل: الانتقال من إحدى الصيغ الثلاث إلى الأخرى لمفهوم واحد رعاية[6] لنكتة.

الوصل: عطف بعض الجمل على بعض.

الفصل: تركه.

الجامع[7] العقلي: أن يكون بين الجملتين اتّحاد في التصوّر.

---

[1] إلى أمر من الأمور: -، آ. [2] مطابقته: مطابقة، آ. [3] لاعتقاد: الاعتقاد، س. [4] له: لها، ب. [5] الكلام الظنّي [...] في المخاطب: -، آ. [6] رعاية: عارية، آ. [7] الجامع: الجمع، آ.

الجامع الوهمي: أن يكون بين تصوّراتهما شبه تماثل.

الجامع الخيالي: أن يكون[1] بين تصوّراتهما تقارن في الخيال سابق لأسباب[2] مؤدّية إلى[3] ذلك.

فصل

الإيجاز: أداء المقصود بأقلّ من عبارة المتعارف.

الإطناب: أداؤه بأكثر منها.

إيجاز القصر: ما لا يكون بحذف.

وقيل: أن يقصر[4] اللفظ على المعنى.

إيجاز الحذف: ما يكون بحذف.

إيجاز التقدير: أن يقدّر معنىً زائد على المنطوق.

التوشيع:[5] أن يؤتى[6] في عجز الكلام بمثنّىً مفسّر باسمين ثانيهما معطوف على الأوّل.

التشبيب: أن يقدّم قبل الشروع ما يمهّد المرام.

الكلام الجامع: أن يحلّي المتكلّم كلامه بشيء من المواعظ[7] أو الحكمة أو شكاية الزمان والإخوان.

إرسال المثل: أن يورد في كلامه مثلاً.

الطرد والعكس: أن يؤتى بكلامين[8] يقرّر الأوّل بمنطوقه مفهوم الثاني، وبالعكس.

التكرير: إعادة الشيء لفائدة.[9]

التفويق: أن يؤتى بمعانٍ ملائمة في جمل مستوية المقدار.

التطرين: أن يؤتى في الكلام مواضع متقابلة كأنّها طراز.

التفسير الخفيّ: أن يؤتى في الكلام لبس فتعمل[10] بما يوضحه.

الإيغال[11]: ختم الكلام أو البيت بما[12] يفيد نكتة يتمّ المعنى بدونها.

---

[1] يكون: تكون، آ س. [2] لأسباب: الأسباب، ب س. [3] إلى: + غير، آ س. [4] يقصر: تقصر، ب. [5] التوشيع: التوشيح، آ. [6] يؤتى: تؤتى، آ س. [7] المواعظ: المواخطع عظ، ب. [8] بكلامين: بكلامه، آ. [9] لفائدة: بفائدة، آ. [10] لبس فتعمل: لما فبعمد، آ؛ لبا فتعمل، ب؛ لبا فيعمد، س. [11] الإيغال: الافعال، آ. [12] بما: -، آ.

التذييل: تعقيب الجملة بجملة تشتمل على معناها للتوكيد.
التكميل: أن يؤتى في كلام يوهم المقصود بما يدفعه.
الاحتراس: مثله.
التتميم: أن يؤتى في كلام لا يوهم خلاف المقصود بفضلة لنكتة.
الترقّي:[1] أن يذكر معنىً ثمّ[2] يردف بما هو أبلغ منه.
الاستطراد: أن يكون في شيء من الفنون ثمّ سنح له فنّ آخر يناسبه فيورده في الذكر.
الاعتراض: أن يؤتى في أثناء الكلام أو بين كلامين متّصلين معنىً بجملة أو أكثر لا محلّ لها من الإعراب لنكتة سوى رفع الإبهام.[3]
الاستتباع: الوصف بشيء يستتبع وصفاً آخر، مدحاً أو ذمّاً.
التغليب: ترجيح أحد المعلومين على الآخر وإطلاق لفظه عليهما.
الإبداع: أن يخترع المتكلّم معانٍ غير مسبوق إليها.
الأسلوب الحكيم: تلقّي المخاطب بغير ما يترقّب وتطلّب السائل بغير ما يتطلّب.

فصل

علم البيان: معرفة إيراد المعنى الواحد في طرق مختلفة بالزيادة في وضوح الدلالة عليه وبالنقصان[4]، ليحترز بالوقوف على ذلك عن الخطأ في مطابقة الكلام لتمام المراد منه.
تمام المراد: إيراد معنىً واحد في عبارة واضحة أو أوضح[5] أو فيما دون ذلك بحسب المقام.
المجاز: الانتقال من الملزوم إلى اللازم.
المجاز المفرد: الكلمة المستعملة في غير ما وضعت له في اصطلاح التخاطب على وجه يصحّ مع قرينة عدم إرادته، وهذا المجاز اللغوي.[6]
المجاز العقلي: الكلام المحكوم فيه بخلاف ما عند المتكلّم بالتأوّل.
الكناية: ترك التصريح بالشيء إلى ما يساويه[7] في اللزوم لينتقل منه إلى[8] الملزوم.

---

[1]الترقّي: التوقّي، ب. [2]ثمّ: -، ب. [3]رفع الإبهام: دفع الابها، ب. [4]بالنقصان: النقصان، ب. [5]أوضح: واضح، آ. [6]اللغوي: -، آ. [7]يساويه: يساريه، ب. [8]إلى: -، س.

وقيل لفظ أريد به لازم[1] معناه مع جواز إرادته معه.

الرمز: ما يشار به إلى المطلوب من قرب[2] مع خفاء.[3]

التلويح: ما يشار به إلى المطلوب من بعد مع خفاء.[4]

الإيماء: هو الكلام المشار به إلى المطلوب من قرب[5] لا مع الخفاء.

التعريض: الكلام المشار به[6] إلى جانب وإيهام أنّ الغرض جانب آخر.

التجريد: أن ينتزع من [أمر] متّصف بصفة[7] [أمر] آخر[8] مثله فيها مبالغة في كمالها فيه.

فصل

التشبيه: وصف الشيء بمشاركته لآخر[9] في معنىً.

الغرض في التشبيه: ما يقصده المتكلّم في إيراد التشبيه.

الأداة:[10] ما يتوصّل به إلى وصف المشبّه[11] بمشاركته المشبّه به في الوجه.

الطرد والعكس: أن يكون الغرض في التشبيه عائداً إلى المشبّه به.

التشبيه المركّب: هو المستعمل فيما شبّه بمعناه الأصلي.[12]

تشبيه التمثيل: للمبالغة.

التمثيل: بمعناه.

المثل: ما ذكر متى فشا استعماله كذلك.

وقيل: هو القول السائر الممثّل مضربه بمورده.

الاستعارة: جعل الشيء الشيء[13] أو للشيء على سبيل المبالغة.

وقيل: هي أن تذكر[14] أحد طرفي التشبيه وتريد[15] به الآخر مدّعياً دخول المشبّه[16] في جنس المشبّه به.

---

[1] لازم: اللازم، آ. [2] قرب: بعد، آس. [3] خفاء: الخفاء، ب. [4] التلويح [...] مع خفاء: -، آس. [5] قرب: قريب، آ. [6] به: -، آ. [7] بصفة: نصفه، آس. [8] آخر: -، ب. [9] لآخر: آخر، ب. [10] الأداة: الإرادة، آ. [11] المشبّه: المشبهة، آس. [12] الأصلي: الأصل، آس. [13] الشيء: -، ب. [14] تذكر: يتدكر، ب. [15] تريد: يريد، ب. [16] المشبّه: الشبه، ب.

الاستعارة الأصلية: أن يكون المستعار اسم الجنس.

الاستعارة التبعية: أن يكون المستعار أفعالاً أو صفاتٍ أو حروفاً.

الاستعارة المطلقة: ما لم يقترن بصفة ولا تفريع.

الاستعارة المجرّدة: ما قرن بما يلائم المستعار له.

الاستعارة المرشّحة: ما قرن بما يلائم المستعار منه.

الاستعارة المصرّح بها: أن يكون المذكور هو المشبّه به.

الاستعارة المكنيّ عنها: أن يكون المذكور هو المشبّه.[1]

الاستعارة التخيّلية: ما لا تحقّق لمعناه حسّاً ولا عقلاً.

المصرّح بها التحقيقية: أن يكون المتروك شيئاً محسوساً.

المصرّح بها التخيّلية:[2] أن يكون المتروك شيئاً متوهّماً محضاً.

المكنيّة: أن يذكر المشبّه ويراد به المشبّه به دالاًّ عليه بقرينة نسبة اللازم المساوي له إليه أو إضافته على سبيل التخيّلية.[3]

التمثيلية: أن يكون الجامع في حكم الواحد.

فصل

علم البديع: ما يعرف به وجوه تحسين[4] الكلام بعد رعاية المطابقة ووضوح الدلالة.

المطابقة: الجمع بين معنيين متقابلين في الجملة.

الطباق والتضادّ: مثلها.

المقابلة: أن يؤتى بمعنيين متوافقين أو أكثر ثمّ يقابل[5] ذلك على الترتيب.

مراعاة النظير: جمع أمر وما يناسبه لا بالتضادّ.

التناسب والتوافق: بمعناها.

تشابه الأطراف: أن يختم الكلام بما يناسب ابتداءاً في المعنى.

---

[1] المشبّه: التشبيح، س. [2] التخيّلية: التخييلية، س. [3] التخيّلية: التخييلية، س. [4] تحسين: التحسين، آ.

[5] يقابل: تقابل، آ ب س.

الإرصاد: أن يؤسّس الكلام على وجه يدلّ على بناء ما بعده.

وقيل: أن يجعل [قبل] العجز من الفقرة أو البيت ما يدلّ عليه إذا عرف الرويّ.

التسهيم: كذلك.

المشاكلة: ذكر الشيء بلفظ غيره لوقوعه في صحبته تحقيقاً أو تقديراً.

المزاوجة: أن يزاوج بين[1] معنيين في الشرط والجزاء.

العكس: أن يقدّم في الكلام جزء ثمّ يؤخّر.

الرجوع: العود إلى الكلام السابق بالنقض لنكتة.

التورية: أن يطلق لفظ له معنيان، قريب وبعيد، ويراد البعيد.

الإيهام: هذه.

الإيهام المجرّد:[2] ما لا يجامع شيئاً مما يلائم القريب.

الاستخدام: أن يراد بلفظ له معنيان أحدهما، ثمّ يضمر الآخر، أو[3] يراد بأحد ضميريه أحدهما، ثمّ بالآخر[4] الآخر.

اللفّ والنشر: ذكر متعدّد على التفصيل أو الإجمال، ثمّ ما لكلّ من غير تعيين، ثقة بأنّ السامع يردّه إليه.

فصل

الجمع: أن يجمع[5] بين متعدّد في حكم.

التفريق: إيقاع تباين بين[6] أمرين من نوع واحد.

التقسيم: ذكر متعدّد ثمّ إضافة ما لكلّ إليه على التعيين.

الجمع مع التفريق: أن يدخل شيئان في معنىً ويفرّق بين جهتي الإدخال.

الجمع مع التقسيم: أن يجمع متعدّد تحت حكم ثمّ يقسّم[7] أو بالعكس.

---

[1]يزاوج بين: في ب عليها خط ترقين وصحّحت في الهامش. [2]المجرّد: المجردة، س. [3]أو: و، آ س. [4]ثمّ بالآخر: -، آ. [5]أن يجمع: يجمع، آ. [6]بين: -، آ. [7]يقسّم: يقيم، آ.

المبالغة: أن يدّعى لوصف بلوغه في الشدّة والضعف بلوغاً مستحيلاً[1] أو مستبعداً لئلا يظنّ أنّه غير متناه[2] فيه.

الإغراق: ما كان الوصف المبالغ فيه ممكناً عقلاً لا عادةً.

المذهب الكلامي: إيراد حجّة المطلوب على طريقة[3] أهل الكلام.

حسن التعليل: أن يدّعى لوصف علّة مناسبة له باعتبار لطيف غير حقيقي.

التفريع: أن يثبت لمتعلّق آخر حكم بعد إثباته لمتعلّق له آخر.

تأكيد المدح بما يشبه الذمّ: أن يستثنى من صفة ذمّ منفية عن [الشيء] صفة مدح بتقدير دخولها فيه أو يثبت لشيء صفة مدح، ويعقّب بأداة[4] استثناء يليها صفة مدح أخرى له.

الاستتباع: المدح بشيء على وجه يستتبع المدح.

الإدماج: أن يضمر كلام سيق[5] لمعنىً آخر.

التوجيه: إيراد الكلام محتملاً لوجهين مختلفين.

تجاهل العارف: سوق المعلوم مساق غيره لنكتة.

القول بالموجب:[6] أن يقع صفة في كلام[7] الغير كناية عن شيء ليثبت له حكم فيثبتها لغيره من غير تعرّض لثبوته له وانتفائه عنه، أو حمل لفظ وقع في كلام الغير على خلاف مراده مما يحتمله[8] بذكر متعلّقه.

الاطّراد: أن يأتي بأسماء الممدوح أو غيره، وآبائه على ترتيب الولادة من غير[9] تكلّف.

فصل

التجنيس: تشابه الكلمتين في اللفظ.

التجنيس التامّ: أن يتّفقا في أنواع الحروف وأعدادها وهيئاتها وترتيبها.

التجنيس المتماثل:[10] ما مرّ، إن كانا من نوع.

---

[1] مستحيلاً: مشتملاً، آ ب س. [2] متناه: ساه، آ ب س. [3] طريقة: طريق، ب. [4] بأداة: بأدائه، آ ب س. [5] سيق: مسبق، آ. [6] بالموجب: الموجب، آ س. [7] كلام: الكلام، آ. [8] يحتمله: يحمله، آ ب س. [9] غير: -، ب. [10] المتماثل: المماثل، آ.

التامّ المستوفى: ما كان من نوعين.

التجنيس الناقص: أن يختلفا في الهيئة دون الصورة.

التجنيس المذيّل:[1] أن يختلفا بزيادة حرف.

التجنيس المضارع: أن يختلفا بحرف أو حرفين مع تقارب المخرج.

المطرّف: بمعناه.

التجنيس اللاحق: أن يختلفا بحرف أو حرفين لا مع التقارب.

تجنيس التصحيف: أن يختلفا بحرفين غير متقاربين اتّفقا كتابة.[2]

المزدوج:[3] ما ولي فيه أحد المتجانسين الآخر.

المردّد[4] **والمكرّر:** مثله.

ردّ العجز على الصدر: هو في الشعر أن يكون إحدى الكلمتين المتكرّرتين والمتجانستين[5] أو الملحقتين[6] بهما في آخر البيت، والأخرى قبلها في أحد المواضع الخمسة من الصدر والحشو والآخر، وفي النثر جعل إحداهما[7] في أوّل الفقرة والأخرى[8] في آخرها.

السجع: تواطؤ الفاصلتين من النثر على حرف واحد.

السجع المطرّف: اختلاف الفاصلتين في الوزن.

السجع المتوازي: ما فيه إحدى القرينتين أو أكثر[9] مثل ما يقابله من الأخرى.

الترصيع: أن يكون الألفاظ مستوية الأوزان متّفقة الأعجاز أو متقاربتها.

التشطير: جعل كلّ شطر من البيت مخالفاً[10] لأخته.

الموازنة: تساوي الفاصلتين في الوزن دون التقفية.

المماثلة: كون ما في إحدى القرينتين أو أكثر[11] مثل ما يقابله في الأخرى من الوزن.

التشريع: بناء البيت على قافيتين يصحّ المعنى على الوقوف على كلّ منهما.

---

[1] أن يختلفا [...] المذيّل: -، س. [2] اتّفقا كتابة: اتفقتا كتبة، آ ب س. [3] المزدوج: المزوّج، آ. [4] المردّد: المردود، آ. [5] المتجانستين: المتجانسين، آ ب س. [6] الملحقتين: المختلفين، آ؛ المختلفتين، ب س. [7] إحداهما: أحدهما، آ ب س. [8] الأخرى: الآخر، آ ب س. [9] أكثر: أكثره، ب. [10] مخالفاً: مخالفة، ب. [11] أكثر: أكثره، ب.

فصل

المقلوب المستوي: ما وقع فيه قلب الكلّ في كلمتين أو أكثر، شعراً أو[1] غير شعر.

المقلوب المجنّح: ما وقع أحد المقلوبين قلب الكلّ في أوّل[2] البيت، والثاني في آخره.

لزوم ما لا يلزم: أن يجئ [قبل] حرف الرويّ أو ما في معناه من الفاصلة[3] ما ليس بلازم في السجع.

الانتحال: أن يؤخذ شعر الغير كلّه من غير تغيير لنظمه.

الفسخ: أن يتّفقا لفظاً ومعنىً[4] قصداً.

الإغارة: أخذ[5] شعر الغير مع تغيير لنظمه أو أخذ بعض اللفظ.

السلخ: مثلها.

الإلمام: أخذ معنى الغير وحده.

المسخ: قلب الصورة الحسنة إلى القبيحة.

المواردة:[6] أن يتّفق المتكلّمان على معنىً من غير أخذ.

الاحتذاء: أن يقتفي الآخر في أسلوب من أساليب البلاغة والفصاحة.

القلب: أن يستعمل معنى الغير في نقيض[7] ما أراده.

الاقتباس: أن يضمّن للكلام شيئاً من القرآن والحديث لا على أنّه منه.

التضمين: أن يضمّن الشعر [شيئاً] من شعر الغير مع التنبيه عليه إن لم يكن مشهوراً عند البلغاء.

الاستعانة: تضمين البيت فما زاد.

الرفو: تضمين المصراع فما دونه.

الإيداع: مثله.

العقد: أن ينظم نثر لا على سبيل الاقتباس.

الحلّ: أن ينثر نظم.

السهل الممتنع: أن يكون الكلام مسبوكاً سبكاً سهلاً وعراً قريباً بعيداً.

التمليح: أن يشار في الكلام إلى قصّة أو شعر من غير ذكره.

[1] أو: و، ب س. [2] أوّل: -، س. [3] الفاصلة: الفاصلتين، آ. [4] معنىً: + و، آ. [5] أخذ: قصد، آ. [6] المواردة: ~~الوارد~~ المواردة، آ. [7] نقيض: نقض، ب.

التلميح: هكذا.

التخلّص:[1] أداء الكلام من نسيب أو غيره إلى المقصود مع رعاية الملائمة[2] بينهما.

الاقتضاب: أداء الكلام من نسيب أو غيره إلى المقصود مع عدم الملائمة[3] بينهما.

فصل الخطاب: تلخيص الكلام إلى المقصود مع قرب الملائمة.[4]

[1] التخلّص: التلخيص، آ ب س. [2] الملائمة: الملامة، آ س؛ الملائكة، ب. [3] الملائمة: الملامة، آ ب س.
[4] الملائمة: الملامة، آ ب س.

الباب العاشر: في العروض

العروض: علم بالقواعد[1] الكلّية المستنبطة من استقراء أشعار العرب التي قالت عليها، ليعرف بها صحيحها[2] من فاسدها.

الشعر: كلام معنوي موزون مقفّى مخيّل بالقصد.[3]

التخيّل: تأثير الكلام في النفس بحيث يحصل في الباطن هيئات مختلفة كالمسرّة والحزن.

الابتداء: اسم لكلّ جزء معتلّ[4] في أوّل البيت بعلّة لا تكون في شيء من الحروف[5] كالخرم.[6]

الاعتماد: اسم للأسباب التي تزاحفها.

الفصل: آخر نصف البيت الأوّل من البيت.[7]

الغاية: آخر نصف البيت الآخر من البيت.[8]

الصحيح: ركن وقع موقع الضرب والعروض وليس فيه نقصان من جهة نقصانات موجبة في بعض الأعاريض والضروب.

الموفور: يقال لركن أوّله وتد ويجوز فيه الخرم[9] ولا يكون فيه.

المتمّم: يقال للركن[10] الثالث في بحر الرمل يكون بلا نقصان كما هو في الأصل.

المعرّى: يقال لركن يكون فيه الإذالة[11] ولا يكون مذالاً.

السبب الخفيف: حرفان أوّلهما متحرّك وثانيهما ساكن.

السبب الثقيل: حرفان متحرّكان.

الوتد المجموع: ثلاثة أحرف تحرّك[12] منها اثنان وسكن[13] الثالث.

---

[1] بالقواعد: بالواعد، آ. [2] صحيحها: صحيحا، ب. [3] بالقصد: بالقصه، ب. [4] معتلّ: تعتل، ب. [5] الحروف: الحرف، ب. وفي الكافي للتبريزي، ص 141: الحشو. [6] الخرم: الجزم، ب. [7] الفصل آخر نصف البيت الأول من البيت: كذا في جميع النسخ. وفي الكافي للتبريزي، ص 141: الفصل كلّ تغيير اختصّ بالعروض ولم يجز مثله في حشو البيت وهذا إنّما يكون بإسقاط حرف متحرّك فصاعداً. [8] الغاية آخر نصف البيت الآخر من البيت: كذا في جميع النسخ. وفي الكافي للتبريزي، ص 142: الغاية كلّ تغيير لزم الضرب مما لا يجوز مثله في الحشو. [9] الخرم: الجزم، ب. [10] للركن: لركن، آس. [11] الإذالة: الإزالة، آس. [12] تحرّك: يحرك، آس. [13] سكن: يسكن، آس.

DOI:10.1163/9789004423367_018

الوتد المفروق: ثلاثة أحرف تحرّك[1] أوّلها وآخرها وسكن[2] أوسطها.
الفاصلة الصغرى: أربعة أحرف تحرّكت[3] الثلاثة الأُول منها وسكن[4] الرابع.[5]
الفاصلة الكبرى: خمسة أحرف تحرّكت الأربعة الأُول وسكن[6] الخامس.[7]

فصل

المقبوض: ما أسقط خامسه الساكن.
المكفوف: ما أسقط سابعه الساكن.
المعاقبة بين الحرفين: أن لا يجوز[8] سقوطهما معاً، وإن جاز ثبوتهما معاً.
المراقبة بين الحرفين: أن لا يجوز سقوطهما جميعاً، ولا ثبوتهما جميعاً.
الخرم: حذف أوّل متحرّك من الوتد المجموع في أوّل البيت.
الخزم: زيادة في أوّل البيت لا يعتدّ بها في التقطيع.
الأثلم[9]: فعولن السالم خاصّة إذا خرم.
السالم: ما سلم من الزِحاف.
المحذوف: ما سقط من آخره[10] سبب خفيف.
المخروم:[11] ما سقط منه جزءان.
المخبون: ما سقط منه [ثانيه] الساكن.
المشكول: ما سقط ثانيه وسابعه الساكنان.
الصدر: ما[12] زوحف لمعاقبة ما قبله فقط.
العجز: ما زوحف لمعاقبة ما بعده فقط.
الطرفان: ما زوحف لمعاقبة ما قبله وما بعده.

---

[1] تحرّك: يحرك، آ س. [2] سكن: يسكن، آ س. [3] تحرّكت: بحركة، آ. [4] سكن: يسكن، آ س. [5] الرابع: الرابعة، ب. [6] سكن: سكنة، آ؛ سكنت، ب س. [7] الخامس: الخامسة، آ ب. [8] لا يجوز: + ~~سقطوه~~، آ. [9] الأثلم: الأثيم، آ؛ الأسلم، ب. [10] آخره: آخر، آ س. [11] المخروم: المجزوم، آ؛ المجزو، س. وفي الكافي في العروض والقوافي، ص 143: المجزوء. [12] ما: -، ب.

البريء: ما سلم من هذه المعاقبة.

المقصور: ما سقط[1] ساكن سببه وسكن متحركه.

المقطوع: ما حذف ساكن وتده وأسكن متحركه.

المطويّ: ما سقط رابعه الساكن.

المخبول: ما سقط ثانيه ورابعه الساكنان.

المذال: ما زيد على اعتداله من عند وتده حرف ساكن.

المعصوب: ما سكن خامسه.

المعقول: ما أسقط خامسه بعد سكونه.

المنقوص: ما أسقط سابعه بعد سكون خامسه.

العضب: خرم[2] مفاعلتن[3] حتّى يصير مفتعلن.

الأقصم: مفاعلتن[4] متى خرم[5] بعد العضب من الوافر حتّى يصير مفعولن.

العقص: خرم[6] مفاعيل[7] من الوافر حتّى يصير مفعول.

الأجمّ: مفاعلن في الوافر إذا خرم حتّى يصير فاعلن.

المقطوف: ما سقط منه زنة[8] سبب خفيف بعد سكون خامسه.

المضمر: ما سكن ثانيه.

الأحذّ: ما أسقط[9] منه الوتد المجموع.

المرفّل: ما زيد على اعتداله سبب خفيف.

الأخرم: مفاعيلن إذا خرم حتّى يصير مفعولن.

الأخرب: مفاعيلن[10] إذا خرم حتّى يصير[11] مفعول.

الأشتر: مفاعلن إذا خرم حتّى يصير فاعلن.

المشطور: ما أسقط منه شطره.

المنهوك: ما أسقط ثلثاه.[12]

---

[1] ما قبله [...] ما سقط: -، س. [2] الخرم: الجزم، ب. [3] مفاعلتن: مفاعيلن، آ ب س. [4] مفاعلتن: مفاعيلن، آ ب س. [5] الخرم: الجزم، ب. [6] الخرم: الجزم، ب. [7] مفاعيل: مفاعيلن، آ ب س. [8] زند: زند، آ؛ زيد، س. [9] الأحذّ ما أسقط: الأخذ ما سقط، آ. [10] مفاعيلن: مفاعيل، ب. [11] يصير: -، س. [12] ثلثاه: ثلثان، ب.

المسبّغ: ما زيد على اعتداله[1] من عند سببه[2] حرف ساكن.[3]
المكشوف: ما حذف متحرك وتده المفروق.
الموقوف: ما سكن متحرك وتده المفروق.
الأصلم: ما سقط وتده المفروق.
المشعّث: ما سقط أحد[4] متحركي وتده.
الأبتر: ما سقط [ساكن] وتده وسكن متحركه وقد سقط من آخره سبب.
المخلّع: جزء[5] أبيات البسيط.

فصل

القافية: هي الكلمة الأخيرة من البيت.
وقيل: هي حرف الرويّ.
وقيل: آخر ساكن في البيت إلى أوّل ساكن يلقاه مع حركة ما قبله.
حرف الرويّ: ما يبنى الشاعر قصيدته عليه.
المجرى: حركة حرف الرويّ إذا كان متحركاً مطلقاً غير مقيّد.
الردف: كلّ ألف أو ياء أو واو ساكنة يكون قبل حرف الرويّ بلا فصل.
الحذْو: حركة ما قبل الردف.
التأسيس: كلّ ألف يدخل بينها[6] وبين حرف الرويّ حرف لا يجب تكريره بعينه.
التوجيه: حركة الحرف الذي يلي الرويّ قبله.
الرسّ: حركة ما قبل التأسيس.
الدخيل: الحرف الداخل بين ألف التأسيس وحرف الرويّ.
الإشباع: حركة الحرف الداخل بين ألف التأسيس وحرف[7] الرويّ.

---

[1] على اعتداله: باعتداله، ب. [2] سببه: سبب، آ س. [3] ساكن: + المفروق، آ ب س. [4] أحد: اخره، آ؛ اخر، س. [5] جزء: جزؤه، آ. [6] بينها: بينهما، آ ب س. [7] حرف: حروف، ب.

الوصل:[1] كلّ ألف أو واو أو ياء أو هاء يكون بعد حرف الرويّ بلا فصل.

الصلة: بمعناه.

النفاذ:[2] حركة هاء[3] الصلة بعد حرف الرويّ إذا كانت متحركة.[4]

الخروج: كلّ ألف أو واو أو ياء يكون بعد الصلة المتحركة.

المتواتر: ما في آخره سبب خفيف.

قيل: ما وقعت حركة بين الساكنين.

المتدارك: ما في آخره وتد مجموع.

قيل: ما توالت فيه حركتان بين ساكنين.

المترادف: ما يجتمع[5] في آخره حرفان ساكنان.

المتراكب: ما في آخره فاصلة صغرى.

قيل: ما توالت في آخره ثلاث حركات.

المتكاوس: ما في آخره فاصلة كبرى.

قيل: ما توالت أربع حركات.

فصل

الإقواء: اختلاف حركة الرويّ في قصيدة واحدة.

الإصراف: كذلك.[6]

الإكفاء: اختلاف حرف الرويّ في قصيدة واحدة.

الإيطاء: أن تتكرّر القافية في قصيدة واحدة بمعنًى واحد.

التضمين: أن تتعلّق قافية البيت الأوّل بالبيت الثاني.

التتميم: مثله.

---

[1] الوصل: الفصل، آ س. [2] النفاذ: النفاء، س. [3] هاء: -، آ. [4] متحركة: صلة، آ. [5] ما يجتمع: لم يجتمع، آ.

[6] كذلك: لذلك، ب.

السناد: أن يأتي بالقافية مرّة مردّفة، ومرّة غير مردّفة في قصيدة واحدة، أو تأتي مؤسّسة مرّة وغير مؤسّسة أخرى، أو يختلف حركة الدخيل بفتحة مع ضمّة أو ضمّة مع كسرة، أو يختلف ما قبل الرويّ المقيّد بفتحة مع ضمّة أو كسرة.

والله أعلم.

الباب الحادي عشر: في المنطق

علم المنطق: آلة قانونية تعصم مراعاتها[1] الذهن عن الخطأ في الفكر.

القانون: أمر كلّي منطبق على جزئيات يتعرّف أحكامها منه.

الفكر: ترتيب أمور معلومة للتأدّي إلى مجهول.

الترتيب: جعل الأشياء المتعدّدة بحيث يطلق عليها اسم الواحد، ويكون لبعضها نسبة إلى بعض بالتقدّم والتأخّر.

البديهي: ما لا يحتاج في تحصيله إلى فكر ونظر.

الضروري: يرادفه.

الكسبي: ما يتوقّف حصوله على الفكر.

النظري: مرادفه.

التصوّر: حصول صورة الشيء في العقل.

التصديق: تصوّر معه حكم.

الحكم: إسناد أمر إلى آخر إيجاباً أو سلباً.

الإيجاب: إيقاع النسبة.[2]

السلب: انتزاعها.

القول الشارح: هو الموصل إلى التصوّر.

الحجّة: هي[3] الموصلة إلى التصديق.

الدلالة: كون الشيء بحالة يلزم من العلم به العلم بشيء آخر.

الوضع: جعل اللفظ بإزاء المعنى أوّلاً.

اللزوم الذهني: كون الأمر الخارجي بحيث يلزم من تحقّق المسمّى في الذهن تحقّقه في الذهن.

اللزوم الخارجي: كونه بحيث يلزم من تحقّق المسمّى في الخارج تحقّقه فيه.

---

[1] مراعاتها: مراتها، آ. [2] النسبة: التثبية، آ. [3] هي: هو، س.

 | DOI:10.1163/9789004423367_019

فصل

المركّب:[1] ما قصد بجزء لفظه الدلالة على جزء معناه.

المركّب الناقص: ما ركّب من اسم وأداة.

المركّب التامّ: ما يدلّ كلّ لفظ منه دلالة تامّة.

المفرد: ما لا يقصد بجزئه الدلالة على جزء معناه.

الأداة: ما لا يصلح للإخبار به أصلاً أو وحده.

الاسم: ما يصلح للإخبار به وحده ولم يدلّ بهيئته[2] وصيغته[3] على زمان معيّن من الأزمنة الثلاثة.

الفعل: ما يصلح للإخبار به وحده ويدلّ بهيئته[4] الوضعية على أحد الأزمنة الثلاثة.

المتواطئ: الاسم الدالّ على معانٍ استوت أفراده الذهنية والخارجية.

المشكّك: كلّ اسم دلّ على معانٍ كان أفراده الذهنية والخارجية[5] فيه بعضها أولى من بعض.[6]

العَلَم: هو الاسم الدالّ على شخص معيّن من الأشخاص.

الجزئي: ما يمنع نفس تصوّر معناه عن وقوع الشركة فيه.

الكلّي: ما لا يمنع عنه.

النوع: المقول على كثيرين متّفقين بالحقائق في جواب "ما هو؟".

الجنس: المقول على كثيرين مختلفين بالحقائق في جواب "ما هو؟".

الفصل: كلّي يحمل على الشيء في جواب[7] "أيّ شيء هو في جوهره؟".

اللازم: ما يمتنع انفكاكه عن الماهية.

العرض المفارق: ما لم يمتنع انفكاكه عن الماهية.

اللازم البيّن: ما يكون تصوره مع تصوّر ملزومه كافياً في جزم الذهن باللزوم بينهما.

اللازم الغير البيّن: ما يفتقر جزم الذهن باللزوم بينهما إلى وسط.

الخاصّة: كلّية مقولة على ما تحت أفراد [حقيقة] واحدة فقط قولاً عرضياً.

العرض العامّ: كلّي مقول على أفراد حقيقة وغيرها قولاً عرضياً.

---

[1] المركّب: + الذي، آ. [2] بهيئته: بهيئة، آ س. [3] وصيغته: وضعية، ب س. [4] بهيئته: بهيئة، آ س. [5] المشكّك [...] والخارجية: في ب كتبت في الهامش. [6] فيه بعضها أولى من بعض: -، آ. [7] جواب: + ما، س.

فصل

القضية: قول يصحّ أن يقال لقائله "إنّه صادق أو كاذب".

الحملية: قضية تنحلّ بطرفيها إلى مفردين.

الشرطية: ما لا تنحلّ[1] بطرفيها إلى مفردين.

المنفصلة:[2] شرطية يحكم فيها بالتنافي بين قضيتين في الصدق والكذب معاً أو في أحدهما فقط.

الموضوع: هو الشيء المحكوم عليه[3] في القضية.

المحمول: هو الشيء[4] المحكوم به فيها.

الرابطة: اللفظة الدالّة على نسبة ارتباط المحمول بالموضوع.

الثلاثية: ما ذكرت فيها الرابطة.

الثنائية: ما[5] حذفت فيها.

المخصوصة: ما يكون الموضوع فيها شخصاً معيّناً.

الشخصية: مرادفها.

المحصورة: مالم يكن الموضوع فيها شخصاً معيّناً، وبيّن فيها كمّية أفراد ما عليه الحكم.

المسوّرة: ترادفها.

السور: اللفظ الدالّ على بيان كمّية أفراد ما عليه الحكم.

الحاصر: يرادفه.[6]

الطبيعية: ما لم يبيّن فيها كمّية أفراد ما عليه الحكم، ولم يصلح أن تصدق كلّية وجزئية.

المهملة: ما لم يبيّن فيها كمّية الأفراد، وتكون صالحة للكلّية والجزئية.

المعدولة:[7] ما يكون حرف السلب فيها جزءاً[8] من الموضوع أو المحمول أو منهما جميعاً.

المحصّلة: موجبة لم يكن حرف السلب فيها جزءاً لا من الموضوع، ولا من المحمول.

البسيطة: سالبة لم يكن حرف السلب فيها جزءاً من الموضوع والمحمول.

مادّة القضية: كيفية نسبة المحمول إلى الموضوع.

---

[1] لا تنحلّ: تنحلّ، آ س. [2] المنفصلة: المتصلة، آ ب س. [3] عليه: -، آ. [4] الشيء: + ~~المحمول~~، آ س. [5] ما: -، آ.
[6] يرادفه: ترادفه، ب. [7] المعدولة: المعدول، آ س. [8] جزءاً: جزء، آ.

جهة[1] القضية: اللفظة الدالّة على كيفية نسبة المحمول إلى الموضوع.

الموجّهة البسيطة: قضية حقيقتها إيجاب فقط أو سلب فقط.

الموجّهة المركّبة: ما تركّبت حقيقتها من إيجاب وسلب.

الضرورية المطلقة: ما يحكم فيها بضرورة ثبوت المحمول للموضوع أو سلبه عنه، ما دام ذات الموضوع موجوداً.

الدائمة المطلقة: ما يحكم فيها بدوام ثبوت المحمول للموضوع، ما دام ذات الموضوع[2] موجوداً.

المشروطة العامّة: ما يحكم فيها بضرورة ثبوت المحمول للموضوع أو سلبه عنه بشرط وصف الموضوع.

العرفية العامّة: ما يحكم فيها بدوام ثبوت المحمول للموضوع أو سلبه عنه بشرط وصف الموضوع.

المطلقة العامّة: ما يحكم فيها بثبوت المحمول للموضوع أو سلبه عنه بالفعل.

الممكنة العامّة: ما يحكم فيها بارتفاع الضرورة المطلقة عن جانب المخالف.

المشروطة الخاصّة: ما يحكم فيها بضرورة ثبوت المحمول للموضوع أو سلبه عنه، ما دام ذات الموضوع موجوداً مقيّداً بقيد اللادوام.

العرفية الخاصّة: ما يحكم فيها بدوام ثبوت المحمول للموضوع أو سلبه عنه بشرط وصف الموضوع مع قيد اللادوام.

الوجودية اللاضرورية: ما يحكم فيها بثبوت[3] المحمول للموضوع أو سلبه عنه بالفعل مع قيد اللاضرورة.[4]

الوجودية اللادائمة: ما يحكم فيها بثبوت[5] المحمول للموضوع أو سلبه عنه بالفعل مع قيد اللادوام بحسب الذات.

الوقتية: ما يحكم فيها بضرورة ثبوت المحمول للموضوع أو سلبه عنه في وقت معيّن من أوقات وجود الموضوع مقيّداً باللادوام بحسب الذات.

المنتشرة: ما يحكم فيها بضرورة ثبوت المحمول للموضوع أو سلبه عنه في وقت غير معيّن من أوقات وجود الموضوع مقيّداً باللادوام بحسب الذات.

الممكنة الخاصّة: ما يحكم فيها بارتفاع الضرورة المطلقة عن جانبي الوجود والعدم جميعاً.

[1] جهة: + جهة، آ. [2] الموضوع: -، س. [3] بثبوت: ثبوت، آ. [4] اللاضرورة: اللاضرورية، س. [5] بثبوت: ثبوت، آ.

فصل

المقدّم: هو الجزء الأوّل من[1] الشرطية.

التالي: هو الجزء الثاني منها.

اللزومية: شرطية متّصلة صدق التالي فيها على تقدير صدق المقدّم لعلاقة بينهما توجب ذلك.

الاتّفاقية: متّصلة يكون صدق التالي فيها[2] على تقدير صدق المقدّم بمجرّد توافق الجزئين على الصدق.

المنفصلة الحقيقية: ما يحكم فيها بالتنافي بين جزئيها صدقاً وكذباً.

المنفصلة المانعة الجمع: ما يحكم فيها بالتنافي بين جزئيها صدقاً فقط.

المنفصلة المانعة الخلوّ: ما يحكم فيها بالتنافي بين جزئيها كذباً فقط.

العنادية: ما يحكم فيها بالتنافي بين جزئيها لذاتي[3] الجزئين.

الاتّفاقية: ما يحكم فيها بالتنافي بين جزئيها لا لذاتي الجزئين بل بمجرّد الاتّفاق.

التناقض: اختلاف القضيتين[4] بالسلب والإيجاب بحيث يقتضي لذاته أن تكون إحداهما صادقة والأخرى كاذبة.

العكس المستوي: جعل الجزء الأوّل من القضية ثانياً، والثاني أوّلاً مع بقاء الصدق والكيفية.

عكس النقيض: جعل الجزء الأوّل من القضية نقيض الثاني، والثاني عين[5] الأوّل مع مخالفته للأصل في الكيفية وموافقته في الصدق.

القياس: قول مؤلّف من قضايا إذا سلّمت لزم عنه لذاته قول آخر.

القياس الاستثنائي: ما كان عين النتيجة أو نقيضها[6] مذكوراً فيه بالفعل.

القياس الاقتراني: ما لم يكن عين النتيجة أو نقيضها[7] مذكوراً فيه بالفعل.

الأصغر: موضوع المطلوب في القياس.

الأكبر: محمول المطلوب في القياس.

الصغرى: هي القضية التي فيها الأصغر.

الكبرى: هي التي فيها الأكبر.

الحدّ الأوسط: ما يكون مكرّراً بين الصغرى والكبرى.

---

[1] من: في، آ ب. [2] فيها: فيهما، آ س. [3] لذاتي: الذاتي، ب. [4] القضيتين: قضيتين، ب. [5] عين: غير، آ س.
[6] نقيضها: بعضها، آ س. [7] نقيضها: بعضها، آ س.

القرينة: اقتران الصغرى والكبرى.

الضرب: يرادفه.

الشكل: الهيئة الحاصلة من كيفية وضع الحدّ الأوسط[1] عند الحدّين الآخرين.

الشكل الأوّل: ما كان الأوسط[2] محمولاً في الصغرى موضوعاً في الكبرى.

الشكل الثاني: ما كان الحدّ الأوسط[3] محمولاً فيهما.

الشكل الثالث: ما كان الأوسط[4] موضوعاً في الصغرى والكبرى.

الشكل الرابع: ما كان الأوسط[5] موضوعاً في الصغرى محمولاً في الكبرى.

فصل

المختلطات: الأقيسة[6] الحاصلة من خلط الموجّهات بعضها مع[7] بعض.

القياس المركّب: تركيب المقدّمات ينتجّ بعضها بنتيجة يلزم منها ومن مقدّمة أخرى نتيجة أخرى إلى أن يحصل[8] المطلوب.

قياس[9] الخلف: إثبات المطلوب بإبطال نقيضه.

الاستقراء: الحكم على كلّي[10] لوجوده في أكثر جزئياته.

التمثيل: إثبات حكم جزئي في جزئي آخر لمعنىً[11] مشترك بينهما.

اليقينيات: قضايا يكون الحكم فيها جازماً مطابقاً ثابتاً.

الأوّليات: قضايا تصوّر طرفيها كافٍ في الجزم بينهما.

المشاهدات: قضايا يحكم بها بقوىً ظاهرة أو باطنة.

المجرّبات: قضايا يحكم بها لمشاهدات متكرّرة مفيدة لليقين.

الحدسيات: قضايا يحكم بها لحدس[12] قويّ في النفس مفيد للعلم.

الحدس: سرعة الانتقال من المبادئ إلى المطالب.

---

[1]الأوسط: الأوسطة، ب. [2]الأوسط: الأوسطة، ب. [3]الأوسط: الأوسطة، ب. [4]الأوسط: الأوسطة، ب. [5]الأوسط: الأوسطة، ب. [6]الأقيسة: الأقية، آ. [7]مع: -، آ. [8]يحصل: يوصل، ب. [9]قياس: + الخط، س. [10]كلّي: كلّ، آب س. [11]لمعنى: لمعان، آس. [12]لحدس: الحدس، ب.

المتواترات:[1] قضايا يحكم بها لكثرة الشهادات بعد العلم بعدم امتناعها والأمن من التواطؤ عليها.

البرهان: القياس المؤلّف من اليقينيات.

البرهان اللمّي: ما يكون الحدّ الأوسط[2] فيه علّة للنسبة في الذهن والعين معاً.

البرهان الإنّي: ما يكون الحدّ الأوسط فيه علّة[3] للنسبة في الذهن فقط.

المشهورات: ما يحكم بها[4] لاعتراف الناس جميعاً بها لمصلحة عامّة أو رقّة أو حمية أو انفعالات من عبادات وشرائع وآداب.

المسلّمات: قضايا يتسلّم من الخصم فيبنى عليها الكلام لدفعه.

الجدل: القياس المؤلّف من المشهورات والمسلّمات.

المقبولات: قضايا تؤخذ ممن يعتقد فيه إمّا لأمر سماوي أو مزيد[5] عقل ودين.

المظنونات: قضايا يحكم بها اتّباعاً للظنّ.

الخطابة: القياس المؤلّف من المقبولات والمظنونات.

المخيّلات: قضايا إذا وردت على النفس أثّرت فيها تأثيراً عجيباً من قبض وبسط.

الشعر: قياس مؤلّف من المخيّلات.

الوهميات: قضايا كاذبة يحكم فيها الوهم في أمور غير محسوسة.

السفسطة: قياس مؤلّف من الوهميات.

المغالطة:[6] قياس يفسد[7] صورته بأن لا يكون على هيئة منتجة لاختلال[8] شرط معتبر.

والله أعلم.

---

[1] المتواترات: المتوازيات، آ. [2] الأوسط: الأوسطة، ب. [3] علّة: -، آ. [4] بها: فيها، س. [5] مزيد: مرند، ب.

[6] المغالطة: المغالطات، س. [7] يفسد: يفيد، آب س. [8] لاختلال: لاخلال، آس؛ الاختلال، ب.

الباب الثاني عشر: في الحكمة

علم الحكمة: ما يبحث فيه عن حقائق الأشياء على ما هي عليه من الوجود بقدر الطاقة [البشرية] ليصير النفس الإنسانية بتحصيلها كاملة مضاهية للعالم العقلي، وتستعدّ بذلك للسعادة[1] القصوى الأخروية.[2]

وقيل: استكمال النفس الإنسانية بالتصوّرات والتصديقات النظرية بقدر الطاقة البشرية.

وقيل: استكمال النفس الإنسانية[3] في قوّتها النظرية بمعرفة الحقّ، وفي قوّتها العملية بفعل الخير.

وقيل: استكمالها بتصوّر الأمور والتصديق بالحقائق النظرية والعملية.[4]

الحكمة النظرية: العلم بما لا يكون وجوده باختيارنا.[5]

الحكمة العملية:[6] العلم بما يكون وجوده باختيارنا.[7]

العلم[8] **الإلهي**: علم ما[9] لا يجب أن يكون في مادّة[10]، ويسمّى العلم الأعلى.

العلم الرياضي: علم ما[11] يجب أن يكون في مادّة [غير] معيّنة، ويسمّى العلم الأوسط.

العلم الطبيعي: علم ما يجب أن يكون في مادّة معيّنة[12]، ويسمّى العلم الأدنى.[13]

الحكمة الخلقية: هي العلم بأحوال تخصّ شخصاً واحداً، ويسمّى علم الأخلاق.

الحكمة المنزلية: هي العلم بأحوال تكون بينه وبين أهله ومنزله، ويسمّى علم تدبير المنزل.[14]

الحكمة المدنية: هي العلم بأحوال تكون بينه وبين عامّة الناس، ويسمّى العلم السياسي.

الفلسفة: هي الحكمة.

وقيل: معرفة الإنسان نفسه.

وقيل: علم الأشياء الأبدية.

العنصر: ابتداء تركيب[15] الشيء.

---

[1] للسعادة: السعادة، آ ب س. [2] الأخروية: الأجزوية، آ. [3] الإنسانية: -، س. [4] العملية: العلمية، آ.

[5] باختيارنا: باختيار، آ. [6] العملية: العلمية، آ. [7] باختيارنا: باختيار، آ. [8] العلم: فالعلم، آ. [9] ما: -، آ س.

[10] مادة: زيادة، آ. [11] ما: -، آ. [12] معيّنة: غير معيّنة، آ ب س. [13] الأدنى: الأذنى، ب. [14] المنزل: المنزلة، آ.

[15] تركيب: تركّب، س.

 | DOI:10.1163/9789004423367_020

الأسطقس: منتهى تحليل الشيء.

الهيولى: القابل للصورة المعيّنة.

المادّة: ما يكون به الشيء موجوداً بالقوّة.

الصورة: ما يكون الشيء[1] به وبغيره موجوداً.

الفاعل: ما يكون في الشيء مؤثّراً.

الغاية: ما يكون مؤثّراً في مؤثّرية الشيء.

الجسم: جوهر ذو أبعاد ثلاثة.

البسيط: جسم لم يتألّف[2] من أجسام مختلفة الطبائع.

المركّب: جسم تألّف[3] من أجسام مختلفة الطبائع.

البسيط العنصري: ما كان جزؤه كالكلّ في الاسم والحدّ.[4]

البسيط الفلكي:[5] بخلافه.

الجماد: جسم مركّب ليس له نموّ.

النبات: جسم مركّب ذو نموّ ليس له[6] حسّ.

الحيوان: حسّاس متحرّك بالإرادة.

فصل

الموجب: هو الذي يجب أن يصدر عنه الأثر.

العلم: حصول[7] صورة الشيء في العقل.

وقيل: تمثّل حقيقة الشيء عند المدرك.

الإدراك: تمثّل حقيقة الشيء عند المدرك بحيث يشاهدها بما[8] به يدرك.

المشاهدة: إدراك الجزئي حاضراً في الخارج.

[1] الشيء: الصورة، آ. [2] لم يتألّف: لم تتألف، ب. [3] تألّف: تألّفت، ب. [4] البسيط العنصري [...] والحدّ: -، ب. [5] الفلكي: الفلك، ب. [6] له: + نموّ، س. [7] حصول: حصوله، آ. [8] بما: ما، آ ب س.

المشاهدة الحقيقية:[1] إدراك عين[2] الحاضر في الخارج.

الحسّ المشترك: قوة إذا ارتسم صورة شيء فيها صار ذلك الشيء مشاهداً.

الخيال: ما يحفظ صور المحسوسات بعد غيبتها.

الواهمة: ما يدرك المعاني الجزئية.

الذاكرة: ما يتذكّرها المدرك.[3]

الحافظة: ما يحفظ المعاني الجزئية.

المتصرّفة: ما يكون بها قوة التركيب والتفصيل بين الصور المأخوذة عن الحسّ، وبين المعاني المدركة بالوهم، وبين الصور والمعاني.

العقل الهيولاني: كون إدراكات النفس بالقوّة.

العقل بالملكة: كون الأوّليات حاصلة لها.

العقل بالفعل: حصول النظريات لها بحيث يقدر على استحضارها متى شاء.[4]

العقل المستفاد: حصول النظريات لها بحيث لا تعزب عنها.

وقيل: حصول النظريات بتلك الحيثية.

الفكر: حركات ما للنفس[5] في المعاني طلباً[6] للحدّ الأوسط، أو ما يجري مجراه.

الحدس: وصول النفس إلى الأوسط بلا فكر.

القوّة الشهوانية: هي الباعثة على جلب المنافع ودفع المضارّ.

القوّة الغضبية: هي الباعثة على الغلبة ودفع المكروه.

النفس الأمّارة: هي القوّة الآمرة[7] للشهوانية والغضبية بالتخيّل والتوهّم إلى جذب اللذّات البدنية ودفع المطالب الحقيقية.[8]

النفس القدسية: هي البالغة في الحدس[9] غايته.

---

[1] الحقيقية: الحقيقة، آ. [2] عين: + عين، آ. [3] المدرك: المدركة، ب. في نسخة ب عبارة "الذاكرة ما يتذكّرها المدركة" تجيء بعد "الحافظة [...]". [4] شاء: شائت، آ س؛ شيات، ب. [5] للنفس: في النفس، آ. [6] طلباً: طلب، آ. [7] الآمرة: الاخرة، ب. [8] الحقيقية: الحقيقة، آ. [9] الحدس: الحدث، آ س.

فصل

الخلاء: أن لا يكون بين الجسمين المتباعدين ما يلاقيانه.

وقيل: هو[1] أبعاد مجرّدة عن المادّة إذا حصلت في مادّة حصل[2] الجسم.

الملأ: جسم من جهة ما تمانع أبعاده دخول جسم آخر فيه.

المكان: هو السطح الباطن من الجسم[3] الحاوي المماسّ للسطح[4] الظاهر[5] من المحويّ.

وقيل: مكان السطح الأسفل الذي يستقرّ عليه شيء ثقيل.

الحركة: هي خروج ما بالقوّة[6] إلى الفعل على التدريج.

وقيل: هي تغيّر أحوال الموجودات.[7]

الكمون: استتار الشيء عن الحسّ.

الكون: حصول[8] ما[9] بالقوّة إلى الفعل دفعة.

التخلخل: ازدياد مقدار الجسم بدون انضمام[10] جسم آخر معه.

التكاثف: انتقاص[11] مقدار الجسم بدون انتقاص[12] جسم عنه.

النموّ: ازدياد جسم باتّصال جسم آخر به مداخلاً في أجزائه في الأقطار على نسبة طبيعية.

الذبول: عكسه.

الحركة: هي الخروج من القوّة إلى الفعل لا في آنٍ واحد.

الحركة السريعة: ما يقطع مسافة أطول في زمان مساوٍ، أو المِثل في زمان أقلّ، أو الأطول في الأقلّ.

الحركة البطئية:[13] عكس ذلك.

الحركة الكوكبية:[14] حركة مركّبة من المستقيمة والمستديرة.

المتحرّك بالذات: ما يكون الحركة حاصلة فيه.

المتحرّك بالعرض:[15] ما يكون الحركة فيما يقارنه.

الحركة القسرية: ما يكون بقوّة حاصلة مستفادة من خارج.

---

[1] هو: هي، آ س. [2] حصل: + لهم، س. [3] الجسم: الجرم، ب. [4] للسطح: + سفل، آ ب س. [5] الظاهر: الظ، س. [6] بالقوّة: + بما، س. [7] الموجود: الموجودة، آ ب. [8] حصول: + ~~انضمام~~، س. [9] ما: -، ب. [10] انضمام: إضمام، آ. [11] انتقاص: انتقاض، آ ب. [12] انتقاص: انتقاض، آ. [13] البطئية: البطعية، ب. [14] الكوكبية: كذا في جميع النسخ. وفي الصحائف للسمرقندي، ص 225: لولبية. [15] بالعرض: بالعروض، ب.

الحركة الإرادية: ما يكون بقوّة حاصلة من خارج مع القصد.
الحركة الطبيعية: ما لا يكون بقصد.
السكون: عدم الحركة فيما من شأنه أن يتحرّك.
الزمان: مقدار حركة الفلك.
الآن: ظرف موهوم يشترك فيه الماضي والمستقبل من الزمان.
وقيل: الزمان صغير المقدار عند الوهم متّصل بالآن الحقيقي من جنبيه.[1]

فصل

القديم بحسب الذات: ما ليس له[2] مبدأ[3] عِلّيّ.
القديم بحسب الزمان: ما ليس وجوده زمانياً.
العلّة الأولى: هي المفردة فوق كلّ صفة عقلية ومنطقية.
الآلة: هي الواسطة بين الفاعل والمنفعل عند وصول أثره إليه.
الفلك: جسم بسيط كري متحرّك بالطبع على الوسط مشتمل عليه.
الملك: جوهر بسيط ذو حياة ونطق هو واسطة بين[4] البارئ والأجسام[5] الأرضية.
الإبداع: إظهار شيء لا من شيء ولا بواسطة شيء.
الخلق: إفادة وجود حاصل عن مادّة وصورة، كيف كان.
الاجتماع: وجود أشياء كثيرة يعمّها معنىً واحد.
الافتراق: مقابله.
المتماسّان[6]: اللذان[7] نهايتاهما معاً في الوضع، وليس يجوز أن يقع بينهما شيء ذو وضع.
التداخل: تلاقي الشيء[8] الآخر بكلّيته حتّى يكفيهما مكان واحد.
المتّصل: ما من شأنه أن يوجد بين أجزائه حدّ مشترك.

[1] جنبيه: جنبية، آ ب. [2] له: + ~~وجوده~~، آ. [3] مبدأ: مبتدأ، آ. [4] بين: من، آ ب س. [5] والأجسام: فالأجسام، آ. [6] المتماسّان: المتجانسان، آ ب. [7] المتجانسان اللذان: المتجانسين اللذين، س. [8] الشيء: شيء، س.

وقيل: القابل للانقسام بغير نهاية.

الاتّحاد: اشتراك أشياء في محمول واحد ذاتي أو عرضي.

وقيل: حصول جسم واحد بالعدد من اجتماع أجسام كثيرة لبطلان خاصّيتها لارتفاع حدودها المشتركة وبطلان نهايتها بالاتّصال.

التتالي: كون الأشياء التي لها وضع ليس بينها شيء[1] آخر من جنسها.

التوالي: كون شيء بعد شيء بالقياس إلى مبدأ محدود وليس بينهما[2] شيء من بابهما.[3]

النهاية: غاية يصير الشيء ذو الكمّية إلى حيث لا يوجد وراءه شيء منه.[4]

الغير المتناهي: كمّ، أيّ أجزاءه[5] أخذت، وجد منه شيء خارج عنه بعينه غير مكرّر.

والله أعلم.

[1] شيء: -، ب. [2] بينهما: بينها، ب. [3] بابهما: انها، آس؛ بابها، ب. [4] منه: عنه، آ. [5] أجزائه: جزئه، س.

الباب الثالث عشر: في الهيئة

علم الهيئة: معرفة تراكيب الأفلاك والأرض.

الفلك المستقيم: دائرة عظمى تحيط بقطبي[1] السماء اللذين عليهما يتحرّك من المشرق إلى المغرب في كلّ يوم وليلة دورة.

معدّل النهار: هو هذا.

خطّ الاستواء: ما يقابل معدّل النهار.

فلك البروج: دائرة ترسمها الشمس بسيرها من المشرق إلى المغرب في السنة.

الدرجة: عبارة عن قسم من ثلاثمائة وستّين قسماً من الفلك إذا قسّم عليها.

الدقيقة: قسم من ستّين قسماً من الدرجة إذا قسّمت عليها.

الثانية والثالثة: على هذا الترتيب.

الأفق: دائرة عظيمة تقطع بسيط الفلك الأعظم، فتفصل بين الظاهر منه فوق الأرض والخفيّ منه تحتها.

دائرة الارتفاع: ما تمرّ بقطبي الأفق.

قوس الارتفاع: قطعة من دائرة الارتفاع.

الميل: بعد الشمس أو الكواكب عن معدّل النهار.

سعة المشرق للشمس من الأفق: ما بين معدّل النهار وبين مطلعها.

عرض البلد: قوس من دائرة نصف النهار الواقع بين القطب والأفق من الطرف الأقرب.

مطالع الفلك المستقيم: ما يطلع مع قسيّ فلك البروج من معدّل النهار في خطّ الاستواء.

مطالع[2] البلد: ما يطلع مع قسيّ فلك البروج من أفق ذلك البلد.

قوس النهار: هي[3] القوس[4] فوق الأرض من الدائرة الموازية لمعدّل النهار التي فيها تدور الشمس.

قوس الليل: ما يبقى من تمام تلك الدائرة.

الجَوزَهَر: النقطتان[5] اللتان يتقاطع عليهما[6] الدائرتان من الأفلاك.

---

[1] بقطبي: بقطب، آ. [2] مطالع: مطلع، آ. [3] هي: هو، س. [4] القوس: + التي، ب. [5] النقطتان: القنطتان، ب. [6] عليهما: -، آ.

 | DOI:10.1163/9789004423367_021

الأوج: أرفع موضع في الفلك الخارج المركز.

الحضيض: أخفض موضع في هذا الفلك وأقرب من الأرض.

منطقة البروج: وسط البروج الذي فيه مسير الشمس.

سير[1] الطول: سير الكواكب[2] في نطاق البروج إلى ما يلي قطب الشمال أو الجنوب.[3]

فلك الأوج: هو الخارج عن المركز.

المركز: سير مركز فلك التدوير في الفلك الخارج المركز.

القبّة: عبارة عن وسط الأرض.

البراكيس:[4] اختلاف المناظر.

المعمورة: ربع الأرض الذي يلي مهبّ[5] الشمال.

فصل

الأزمان:[6] هي أجزاء الساعات المعوجّة.

الساعة المعوجّة: هي نصف سدس النهار أو الليل الذي ليس بمعدّل.

الساعة الزمانية: كذلك.

الساعة المستوية: مقدار ما يدور من الفلك خمس عشرة[7] درجة.

وسط الكوكب: سيره[8] الوسط في فلكه[9] الخاصّ[10] الخارج المركز.

السير المعدّل: تقويمه وهو حركته في فلك البروج.

الخاصّة: سير الكوكب نفسه في فلك التدوير.

رجوع الكوكب: سيره[11] طولاً على خلاف نضد البروج.

الرجعة: بهذا المعنى.

---

[1] سير: + الشمس، آ. [2] الكواكب: الكوكب، ب. [3] الجنوب: لجنوب، آ. [4] البراكيس: كذا في جميع النسخ، و في مفاتيح العلوم للخوارزمي: البركسيس. [5] مهبّ: مهبت، آ. [6] الأزمان: الزمان، ب. [7] خمس عشرة: خمسة عشر، آ ب س. [8] سيره: سيرة، آ ب س. [9] فلكه: فلک، س. [10] الخاصّ: -، آ س.
[11] سيره: شيره، آ.

الكبد: نجم في السماء لا يرى.

البهت:[1] سير[2] الكوكب المعدّل ليوم وليلة.

النهندر:[3] ما يبقى من مسير الكوكب ليوم وليلة.

الكوكب الصميمي:[4] ما يكون بينه وبين الشمس ست عشرة[5] دقيقة فما دونها.

الاحتراق: أن يكون الكوكب مقارناً للشمس وبينهما أكثر من دقائق.

الصميم تحت الشعاع: أن يكون الكوكب مع الشمس قبل الاحتراق أو بعده.

الكبيسة: ما يجتمع[6] من ربع يوم في كلّ سنة، فإذا مضت أربع[7] سنين انجبرت[8] الأرباع فصارت يوماً واحداً.

والله أعلم.

---

[1]: البهت: النهب، آ ب س. [2] سير: مسير، ب. [3]: النهندر: البهبندر، آ؛ البهيندر، ب؛ البهنيدر، س.
[4] الصميمي: الضممي، آ؛ الضميمي، س. [5] ستّ عشرة: ستّة عشر، آ ب س. [6] ما يجتمع: -، ب. [7] أربع: ربع، آ. [8] انجبرت: الجيرت، آ ب.

الباب الرابع عشر: في الهندسة

الهندسة: علم يعرف به مقادير الأشياء.

أقليدس: اسم رجل صنّف كتاب **الأسطقسّات** وجمع فيه أصول الهندسة.

المصادرة: ما يصدّر الكتاب أو الباب[1] به من أبواب الهندسة.

المقدار: ما له قدر جسماً أو بسيطاً[2] أو خطّاً أو زاوية أو غيرها.

وقيل: كلّ ذي[3] بعد من الخطوط والبسائط والأجسام.

النسبة: إضافة ما في القدر بين مقدارين متجانسين.

المقادير المتجانسة: ما يمكن إذا ضوعف أن يزيد بعضها على بعض.

تبديل النسبة: أخذ المقدّم عند المقدّم[4] والتالي عند التالي.

تركيب النسبة: أخذ المقدّم مع التالي بمنزلة شيء واحد عند التالي.

تفضيل النسبة: أخذ فضل المقدّم على التالي عند التالي.

قلب[5] النسبة: أخذ المقدّم عند فضله على التالي.

نسبة المساواة: أن يفضل كمّية المقادير على عدّتها فتنسب الأطراف بعضها ببعض.

الأبعاد: هي الطول والعرض والعمق أو السمك.

البسيط: ما له طول وعرض فقط وأطرافه خطوط.

الخطّ: ما له طول فقط ونهايته نقطتان.

النقطة: شيء ذو وضع لا جزء له.

الخطّ المستقيم: هو الموضوع على مقابلة أيّ النقط[6] كانت عليه، بعضها لبعض.

البسيط المستوي:[7] هو الموضوع على مقابلة أيّ[8] الخطوط المستقيمة كانت عليه، بعضها لبعض.

السطح: كذلك، ولا يدرك بالحسّ إلّا مع الجسم.

الخطوط المتلاقية: هي التي تلتقي وتحيط بزاوية.

---

[1] الباب: النبات، آ. [2] بسيطاً: بسطاً، آ. [3] ذي: ذو، ب. [4] عند المقدّم: -، آ. [5] قلب: + ~~التالي~~، س.

[6] النقط: النقطة، آ ب س. [7] المستوي: المسوي، س. [8] أيّ: + ~~الموضو~~، آ.

 | DOI:10.1163/9789004423367_022

العمود: خطّ مستقيم قائم على خطّ مستقيم كانت الزاويتان عن جنبيه متساويتين.[1]

الزاوية القائمة: إحدى جنبتي[2] العمود القائم على الخطّ المستقيم.

الزاوية المنفرجة: أكبر إحدى جنبتي[3] خطّ مستقيم، قائم على خطّ مستقيم.[4]

الزاوية الحادّة: أصغرها.

الزوايا المسطّحة: ما يحدث عن التقاء ثلاثة خطوط على غير استقامة.

الزوايا المجسّمة:[5] ما يحدث عن التقائها على استقامة.

الساقان: الخطّان اللذان يحيطان بزاوية.

القاعدة: خطّ يصل بين طرفي الساقين.

المتوازيان: خطّان أخرجا على سطح واحد إذا أخرجا بالاستقامة لم يتلاقيا.

فصل

الدائرة: شكل مسطّح يحيط به خطّ واحد في داخله نقطة، كلّ الخطوط المستقيمة الخارجة[6] منها إلى الدائرة متساوية.

القطر: خطّ يخرج من طرف زاوية وينتهي إلى أخرى.

قطر الدائرة: خطّ مستقيم يمرّ بمركز من المحيط إلى المحيط.

نصف الدائرة: شكل يحيط به القطر.

القوس: التي حازها[7] القطر[8] من الخطّ المحيط.

الوتر: خطّ يصل بين طرفي المقوّس أو المنحنى.

قطعة الدائرة: شكل يحيط به خطّ مستقيم، وقوس من الخطّ المحيط بالدائرة، إمّا أكبر من نصفه أو أصغر.

وتر القطعة: هو قاعدة القطعة.

---

[1] متساويتين: متساويين، س. [2] جنبتي: جنبي، آ. [3] جنبتي: جنبي، آ. [4] قائم على خطّ مستقيم: -، آ س.
[5] المجسّمة: المسجية، آ. [6] الخارجة: الخارجية، آ. [7] حازها: خانها، آ. [8] القطر: الطقر، ب.

السهم: خطّ[1] يخرج[2] من النقطة[3] التي تقسم وتر القوس بنصفين ويحيط[4] مع الوتر بزاوية قائمة.

الجيب المستوي: هو نصف الوتر.

الجيب المعكوس: خطّ يقطع كلّ واحد من القوس ووترها بنصفين.

جيب تمام القوس: جيب ينقطع[5] عن ربع دائرة.

وتر تمام القوس: وتر ما ينقص القوس عن[6] نصف دائرة.

الجيب الأعظم: هو نصف قطر الدائرة.

الدوائر المتساوية: هي التي أقطارها متساوية.

الخطّ المماس للدائرة: ما يلقاها ولا يقطعها إذا أخرج في كلتا[7] الجهتين.

الدوائر المماسّة: هي التي تتلاقى ولا تتقاطع.

الزاوية التي في قطعة الدائرة: ما[8] إذا أعلم على قوس القطعة علامة، وأخرج منها إلى نهايتي قاعدة القطعة خطّان أحاط[9] بها.

زاوية[10] القطعة: ما يحيط بها قوس القطعة ووترها.

قطع من الدوائر المتشابهة: هي التي زواياها متساوية أو تقبل زوايا[11] متساوية.

فصل[12]

الشكل: ما يحيط به[13] حدّ أو حدود.

الشكل المستقيم: ما يحيط به[14] خطوط مستقيمة.

المربّع: ما يحيط به أربعة خطوط.

الصحيح:[15] قائم الزوايا متساوي الأضلاع.

المستطيل: قائم الزوايا متساوى كلّ ضلعين متقابلين.

---

[1] خطّ: -، آ. [2] يخرج: يجمع، آ س. [3] النقطة: النقط، ب. [4] ويحيط: أو يحيط، آ. [5] ينقطع: يقطع، آ. [6] عن: عند، ب. [7] كلتا: كلّي، آ؛ كلتي، ب س. [8] ما: -، س. [9] أحاط: أحاطا، ب. [10] زاوية: -، ب. [11] زوايا: زواياها، آ س. [12] فصل: -، آ. [13] به: + خطّ، س. [14] به: -، س. [15] الصحيح: الصحيحة، آ ب س.

المعيّن: المتساوي الأضلاع غير قائم الزوايا متساوي كلّ ضلعين متقابلين.
المحرّف: ما كان خارجاً من هذه الحدود.
السطوح المتشابهة: ما تساوت زواياها وتناسبت الأضلاع المحيطة بزواياها المتساوية.
السطوح المتكافئة الأضلاع: ما تناسبت أضلاعها على التقديم والتأخير.
السطح الهلالي: ما يحيط به خطّان مقوّسان أحدهما إلى أخمص الآخر.
السطح البيضي: ما يحيط به قوسان متقابلا[1] الأخمصين.
الشكل القطّاع: قطعة دائرة رأسها إمّا على مركزها[2] وإمّا على محيطها مثل هذين الشكلين:[3]

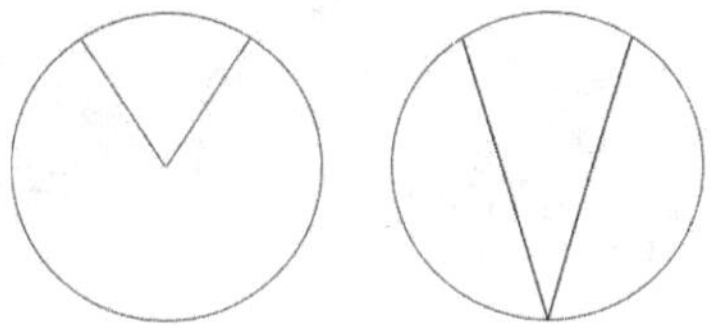

البسيط المقبّب الكري: ما كان على شكل الكرة.
البسيط الأسطواني: ما يبتدئ من دائرة و ينتهي إلى دائرة.
البسيط المقبّب المخروط: ما يبتدئ من نقطة و ينتهي إلى محيط دائرة.
الشكل الصنوبري: يرادفه.
الشكل الناري: جسم تحيط به أربعة سطوح مثلّثات متساوية[4] الأضلاع.
الشكل الهوايي:[5] جسم تحيط به ثمانية سطوح مثلّثات متساوية الأضلاع والزوايا.
[**الشكل الأرضي**: جسم تحيط به ستّة سطوح مربّعات متساوية الأضلاع والزوايا.][6]
الشكل المكعّب: يرادفه.[7]
الشكل المائي: جسم يحيط به عشرون مثلّثاً متساوية الأضلاع والزوايا.

[1] متقابلا: متقابلاً، آ؛ متقابلان، س. [2] مركزها: ما ذكرنا، آ س؛ ذكرنا، ب. [3] الشكلان ساقطان من النسخ الثلاث، و رسمنا إيّاهما مبنيّ على ما يقول المؤلّف في تعريف الشكل القطّاع. [4] متساوية: متساويات، آ.
[5] الهوايي: الأرضي، آ ب س. [6] [الشكل الأرضي: ... والزوايا.]: أضفناه من مفاتيح العلوم، ص 188، ليستقيم ما يقول المصنّف فيما بعده: "الشكل المركّب يرادفه". [7] يرادفه: يرادفانه، آ.

الشكل الفلكي: جسم يحيط به اثنى عشر سطحاً مخمّساً[1] متساوية الأضلاع والزوايا.[2]

الشكل اللبني: جسم مربّع يكون بُعدان[3] من أبعاده متساويين والثالث أصغر من شكل المربّعة.

الشكل العمودي: مربّع يكون بُعدان من أبعاده متساويين والثالث أعظم.

الشكل البئري:[4] هو العمودي بعينه في تسمية بعضهم.

الشكل اللوحي: مربّع مختلف أبعاده الثلاثة.

الارتفاع في الشكل:[5] هو العمود[6] المخرج من رأسه إلى خطّ قاعدته.

الجسم المنشور: ما يحدث عن أحد الأجسام المربّعة إذا قسّم بنصفين على أحد أقطاره.

الجسم المخروط: شكل يبتدئ من نقطة وينتهى إلى محيط دائرة يحيط به بسيط صنوبري ودائرة.

الكرة: شكل مجسّم يحدث عن إثبات قطر نصف الدائرة وإدارة نصف[7] الدائرة إلى أن يعود إلى حيث بدأ بالحركة.

وقيل: جسم يحيط به[8] بسيط واحد في داخله نقطة، كلّ الخطوط [المستقيمة] الخارجة من تلك النقطة التي بسيطها متساوية، وتلك النقطة مركزها، وينتهي إلى بسيطها.

بسيط الكرة: ما يحدث عن قوس نصف الدائرة.

محور الكرة: قطر نصف الدائرة المثبت.

وقيل: قطرها الذي يتحرّك عليه وهو ثابت.

قطب الكرة: طرفا نصف الدائرة المثبت.

الدوائر العظام: هي التي على بسيط الكرة، وتمرّ بمركز الكرة.

البيضية: شكل مجسّم يحيط به بسيط واحد ويحدث عن قطعة أقلّ[9] من نصف دائرة إذا صيّر[10] طرفها كالمحور، وأدير[11] إلى أن يرجع إلى حيث ابتدأ منه.

الحلقة:[12] جسم يحيط به بسيط واحد مستدير[13] في داخله مكان يمكن أن يقع فيه كرة.

الأسطوانة: جسم يبتدئ من دائرة وينتهى إلى دائرة مساوية لما يحيط به بسيط[14] أسطوانيّ.

والله أعلم.

---

1 مخمّساً: مجسماً، آ س. 2 الشكل الفلكي [...] والزوايا: -، ب. 3 بعدان: بعد، آ ب س. 4 البيري: البيروي، آ. 5 الشكل: -، آ. 6 العمود: العمودي، آ س. 7 نصف: نصفه، س. 8 به: -، آ. 9 أقلّ: اول، آ ب س. 10 صيّر: طير، آ ب س. 11 أدير: أريد، ب. 12 الحلقة: الخلقة، آ. 13 مستدير: مستديرة، آ. 14 بسيط: + أو، آ.

الباب الخامس عشر: في الحساب

علم الحساب: تتبّع كيفية استخراج المجهولات من المعلومات والتوصّل إليها.[1]

الوحدة: ما بها يقال لكلّ موجود "واحد".

العدد: جماعة مركّبة من آحاد.

وقيل: نصف[2] مجموع الحاشيتين.

العدد الأوّل:[3] ما يعدّه الواحد فقط.

وقيل: ما ليس له مقدار يعدّه غير الواحد.

العدد المركّب: ما يعدّه عدد بعد عدد[4] آخر مثله أو مخالفاً له.[5]

العدد المجسّم: ما يعدّه عدد بعد عدد[6] مركّب متساوية الأبعاد أو مختلفتها.

العدد التامّ: ما إذا جمعت أجزاءه الصحيحة كانت مثله.[7]

العدد الزائد: ما إذا جمعت أجزاءه كانت أكثر منه.

العدد الناقص: ما إذا جمعت أجزاءه كانت أقلّ منه.

العددان المتجانسان: ما يكون أجزاء كلّ واحد منهما مثل الآخر.

الأعداد المتناسبة: ما يكون نسبة الأوّل منه إلى الثاني كنسبة الثالث إلى الرابع.

الأعداد المجسّمة المتشابهة: ما لا يكون أضلاعها متناسبة.

الأعداد المسطّحة المتشابهة: يكون[8] كذلك.

الأعداد النظيرة في النسبة: هي المقدّمات للمقدّمات والتوالي للتوالي.

العدد الهوهوي: المجسّمات[9] إذا كان سمك أحدها مثل ضلع من أضلاعه.

العدد الدوائري: ما كان بدؤه ونهايته شيئاً واحداً.

---

[1] إليها: إليه، آ ب. [2] نصف: نصفه، آ س. [3] الأوّل: أول، ب. [4] بعد عدد: بعدد، ب. [5] له: -، آ س.
[6] بعد عدد: بعدد، ب. [7] مثله: مثاله، آ ب. [8] يكون: -، ب. [9] المجسّمات: المخمسات، ب س.

 | DOI:10.1163/9789004423367_023

فصل

الفرد: ما لا ينقسم قسمين متساويين.

وقيل: ما ينقسم قسمين غير متجانسين.

وقيل: هو الذي بينه وبين الزوج واحد[1] فقط.

الزوج: ما ينقسم قسمين متساويين.

وقيل: متجانسين.

زوج الزوج:[2] ما يعدّه عدد زوج بعدد زوج فقط.

زوج الفرد: ما لا ينتصف إلّا مرّة واحدة.

فرد الفرد: ما يعدّه عدد فرد بعدد فرد.

زوج الزوج والفرد: ما ينتصف أكثر من مرّة واحدة.

النسبة: أن ينسب العدد إلى آخر، فيقال: نصفه أو ثلثه.

العيار: تشبيه النسب.

إبدال النسبة: هو النسبة للمقدّم إلى المقدّم، والتالي إلى التالي.

تركيب النسبة: نسبة مجموع المقدّم والتالي إلى التالي أو إلى المقدّم.

تفضيل النسبة: نسبة الفضل بين المقدّم والتالي إلى التالي أو إلى المقدّم.

عكس النسبة: نسبة التالي إلى[3] المقدّم.

قلب النسبة: نسبة المقدّم إلى الفضل بينه وبين التالي.

الجذور: الأعداد التي يعبّر بها عن النسب.

المال: كلّ ما يجتمع من ضرب عدد في نفسه.

الجذر: كلّ ما يضرب في نفسه.

الجذر المطلق: ما يعرف حقيقة مقداره ويمكن أن ينطق به.

الجذر الأصمّ: ما لا سبيل إلى علم حقيقته بالعدد.

مثلّثة[4] القواعد: ما يتولّد من تراكم المثلّثات.

مربّعة القواعد: ما يتولّد من تراكم المربّعات.

---

[1] واحد: أحد، آ س. [2] الزوج: الفرد، س. [3] إلى: -، ب. [4] مثلّثة: + الهوا، آ.

ذو الاسمين: ما لا[1] يمكن أن ينطق [به] بلفظ واحد.

المكعّب: هو المال إذا ضرب في جذره، فالمبلغ هو المكعّب، وذلك الجذر هو الكعب.

مال المال: ما إذا[2] اجتمع من ضرب المال في نفسه[3]، وكذلك إذا ضرب المكعّب[4] في كعبه.

المدار: مبلغ يحصل من ضرب[5] مال المال في المكعّب.

الشيء:[6] في كلام أهل الجبر هو الجذر المجهول.

القسمة: أخذ حصّة الواحد من المقسوم عليه.

والله أعلم.

---

[1] لا: -، آ س. [2] إذا: -، ب س. [3] في نفسه: -، آ. [4] المكعّب: الكعب، آ ب س. [5] ضرب: -، آ س.
[6] الشيء: الخير، ب.

الباب السادس عشر: في الاستيفاء

الاستيفاء: علم بقوانين يعرف بها ضبط مداخل أموال الديوان وإخراجاتها لكيفيات[1] المحاسبات وكمّياتها.

قانون الخراج: أصله الذي يرجع إليه ويبنى الجباية عليه.

الأوارج: ما ينقل من القانون على إنسان إنسان ويثبت ما يؤدّيه دفعة بعد أخرى إلى أن يستوفى ما عليه.

الروزنامج: كتاب اليوم يثبت فيه ما يجري من استخراج أو نفقة.[2]

الختمة: حساب يرفعه الجهبذ في كلّ شهر بالاستخراج والجمل والنفقات.

الختمة الجامعة: ما يعمل هكذا لكلّ[3] سنة.

التأريج:[4] سواد يعمل [للعقد] لعدّة أبواب يحتاج إلى علم جملها.

العريضة: شبيهة بالتأريج[5] إلّا أنّها لأبواب يحتاج أن يعلم فصل ما بينهما فينقص[6] الأوّل من الأكثر من بابين منها، ويوضع ما يفصل في باب واحد، والمقصود من العريضة هو.

البراءة: حجّة يبذلها الجهبذ أو الخازن للمؤدّي ما يؤدّى إليه.

الموافقة: حساب جامع يرفعه العامل عند فراغه من العمل باتّفاق بين الرافع والمرفوع[7] [إليه].

الجماعة: هي الموافقة.

المحاسبة: ما تفرّد بها الرافع و[8] المرفوع [إليه] على تفصيلاته.

الجريدة: ما يكتب[9] في كلّ سنة بأسامي الرجال وكناهم وأنسابهم[10] وحلاهم ومبالغ أرزاقهم وقبوضهم.

الرجعة: حساب يرفعه[11] المعطى[12] في بعض العساكر.

---

[1] الكيفيات: لكيقيابات، آ؛ لكيفياتهايات، س. [2] نفقة: نقطة، ب. [3] لكلّ: الكل، ب. [4] التأريج: التاريخ، آ س. [5] بالتأريج: بالتاريخ، س. [6] فينقص: فينقض، آ. [7] المرفوع: + ~~على تفصيلاته~~، س. [8] و: أو، ب. [9] يكتب: كسر، آ؛ يكسر، ب س. [10] وأنسابهم: أو نسابهم، ب. [11] يرفعه: يرفع، آ ب س. [12] المعطى: + ما، آ س.

 | DOI:10.1163/9789004423367_024

الرجعة الجامعة: ما يرفعها صاحب ديوان الجيش لكلّ جمع من صنوف الأرزاق.

الصكّ: ما يعمل لكلّ طمع يجمع فيه أسامي المستحقّين وعددهم ومبلغ مالهم ليوقّع السلطان عليه بإطلاق الرزق عليهم.

المؤامرة: ما يجمع فيه الأوامر الخارجة في مدّة أيّام الطمع ليوقّع السلطان فيه إلى استئمار واستدعاء توقيع.

الاستقرار: ما يعمل لما استقرّ عليه أمر الطمع بعد الإثبات والفكّ والوضع والزيادة والحطّ والتحويل والنقل وغيرها.

السجّل: ما يكتب لرسول أو لمجمّز[1] أو رحّال بإطلاق نفقته حيث بلغ إلى كلّ عامل يجتاز به، ويقال أيضاً لما يكتبه القاضي لتفصيل[2] القضاء.

الفهرست: ذكر الأعمال والدفاتر التي تكون في الديوان.

الدستور: نسخة الجماعة المنقولة من السواد.

الترقين: ما يخطّ في التأريج[3] أو العريضة إذا خلا باب من السطر[4] ليكون الترتيب محفوظاً.

الجائزة: علامة المقابلة.

فصل

الخراج: ما يؤخذ من الصلح.

العشر: ما يؤخذ من الأرض التي أسلم أهلها أو أحياها المسلمون، أو القطائع.

الحشري: ميراث من لا وارث له.

مال الجوالي: ما يؤخذ عمّن جلا عن وطنه.

مال الجماجم: كذلك.

المكس:[5] ضريبة تؤخذ من التجّار في المراصد.

الطسق:[6] وضيعة[7] تؤخذ على كلّ صنف من الزروع لكلّ جريب قدر معلوم.

---

[1] لمجمّز: لمحمّن، آ؛ لمحمز، س. [2] لتفصيل: لفصل، ب. [3] التأريج: التاريخ، س. [4] السطر: السفر، آ ب س.

[5] المكس: المكسر، آ. [6] الطسق: السطق، آ. [7] وضيعة: وظيعة، ب.

الإقطاع: أن يعطي السلطان أرضاً لرجل[1] فيصير له رقبتها ويؤدّي عشرها ويكون لورثته بعده.

القطائع: تلك الأراضي.

الطعمة: أن يدفع الضيعة ليعمّرها ويؤدّي عشرها مدّة حياته[2]، فإذا مات ارتجعت عن[3] ورثته.

الحماية: أن يحمي ضيعة فلا يدخلها عامل ويوضع عليها شيء يؤدّى في السنة إلى بيت المال، وقد يقال لها: الإيغار.

الإيغار: تحويل الخراج من بلد إلى بلد آخر.

التسويغ: أن يسوّغ شيء من الخراج في السنة.

الحطيطة والبريكة: كذلك.

التقرير: أن يقرّر العامل القوم بالبقايا حتّى إذا أقرّوا سقط[4] ذكر القوم.

الحاصل: ما يكون في بيت المال أو كان باقياً على العامل أو الرعية[5] لم يستخرج بعد منهم.

العبرة: ثبت الصدقات لكلّ كورة بأن يجمع ارتفاع السنة الكثيرة الريع والسنة قليلة الريع فيؤخذ نصفهما[6] بعد اعتبار الأسعار[7] وسائر العوارض الواقعة.

الراتبة: النفقات التي لا بدّ منها.

العارضة:[8] نفقات تحدث.

الرائج:[9] مال يسهل استخراجه.

المنكسر:[10] ما لا مطمع في استخراجه.

المردود:[11] ما يردّ عليه ولا يحسب[12] له.

الموقوف: ما ينتظر استئمار السلطان في حبسه[13] أو ردّه.

الحزر: تقدير غلّات[14] الزروع.

الخرص: تقدير النخل والكرم خاصّة.

التخمين: تقدير الخضروات.[15]

---

[1] لرجل: -، آ س. [2] حياته: حيوة، آ. [3] عن: من، آ ب. [4] أقرّوا سقط: أفرد أسقط، آ ب س. [5] أو الرعية: والرعية، آ س؛ أو لرعية، ب. [6] نصفهما: نصفها، ب. [7] الأسعار: الأسغار، آ؛ الأسفار، س. [8] العارضة: العارضية، آ. [9] الرائج: الرابع، آ. [10] المنكسر: المنكس، ب. [11] المردود: المرود، ب. [12] لا يحسب: لا يجنب، آ. [13] حبسه: جنسه، آ س. [14] غلّات: غلاف، آ؛ غلاق، س. [15] الخضروات: الخضراوات، ب.

التلجئة: تسليم الضعيف ضيعة إلى قويّ ليحامي عليها.

المصادرة والمصالحة: بمعنى المواقفة.

والله أعلم.

الباب السابع عشر: في الموسيقى

الموسيقى: علم يعرف بها كيفية تأليف الألحان وكيفية الإيقاعات.

التأليف: تركيب النغمات.

النغمة: صوت غير متغيّر إلى حدّة.

الحدّة والثقل: كيفيتان مسموعتان تعرضان لبعض الأصوات بالقياس إلى آخر.

الوزن: الأزمنة التي تقع بين النغمات على وجه يحكم بصحّته واستقامته الطبع المستقيم.

البعد: صوت بين النغمتين يبتدأ فيه بنغمة، ويثنّى فيه بنغمة أخرى.

البعد المتّفق: نغمتان مركّبتان تميل اليهما النفس وتجد منهما[1] لذّة.

البعد المتنافر: نغمتان مركّبتان تكرههما النفس.

الجماعة: تأليف النغمات على وجه يلائم.

وقيل: نغمات يولد[2] منها لحن.

الطبقات: مراتب حدّة الصوت أو ثقله.

البعد ذو[3] الخمس: هو مثل ما بين مطلق البم إلى سبّابة المثلث.

البعد ذو الأربع: هو ما بين مطلق البم إلى خنصره، وهو ربع[4] الوتر.

البعد الطنيني: هو ما بين المطلق والسبّابة.

المدّة والعودة: كذلك.

الفضلة: بعد ما بين البنصر والخنصر، أو ما بين السبّابة ووسطى الفرس.

وقيل: هي نصف المدّة بالتقريب و[5] البقية مثلها.

الإرخاء: نصف[6] الفضلة.

الخطّ: نغم مطلق البم عند نغمة سبّابة المثنى على التسوية المشهورة.

---

[1] تجد منهما: لا تجد منها، آس؛ تجد منها، ب. [2] نغمات يولد: نغمتان تولد، آ. [3] ذو: دون، ب. [4] ربع: رابع، آ. [5] و: -، آس. [6] نصف: + ا نصف، س.

فصل

الدستان: اسم لكلّ من الألحان المنسوبة إلى فاربذ.

وقيل: علامات ترسم على ذوات الأوتار يعرف بها متى أريد[1] إخراج نغمة من أيّ جزء من أجزاء[2] الوتر تخرج.

وقيل: هي الرباطات التي يوضع عليها الأصابع.

مشط العود: مسطرة تشدّ عليها الأوتار.[3]

عينا العود: الثقبتان اللتان على وجهه.

الإ بريق: عنق العود بما فيه من الآلات.

المضراب: ما يضرب به الأوتار.

الجسّ: نقر الأوتار بالسبّابة والإبهام دون المضراب.

الحزق: مدّ الوتر.

الإرخاء: ضدّه.

البمّ: أغلظ الأوتار[4] الأربعة من العود.

المثلث: الذي يليه.

المثنيّ: الذي يلي المثلث.

الزير: أدقّ الأربعة.

الزائد: الدستان الأوّل الذي بإزاء ب من العود.

المجتثّ: الدستان الثاني الذي بإزاء ج منه.

دستان السبّابة: هو الثالث الذي بإزاء د منه.

الدستان الوسطى الفرس: هو الرابع الذي بإزاء ه منه.

دستان الوسطى[5] زلزل: هو الذي بإزاء و منه.

دستان البنصر: هو الذي بإزاء ز[6] منه.

دستان الخنصر: هو الذي بإزاء ح منه.

---

[1]أريد: -، ب. [2]أجزاء: الأجزاء، ب. [3]الأوتار: الأوتا، آ. [4]الأوتار: اوالاوتار، ب. [5]الوسطى: وسطى، آس. [6]ز: -، آ.

فصل

الإيقاعات: نقرات يقع بينها[1] أزمنة محدودة تشتمل على أدوار متساوية الكمّية على أوضاع[2] مخصوصة يدرك تساوي تلك الأدوار والأزمنة بالطباع[3] المستقيمة.

الثقيل الأوّل: أدوار يمكن أن يتلفّظ في زمان كلّ دور منها بثمانية أسباب ثقال.

الثقيل الثاني: ما يساوي زمان كلّ دور منها زمان الثقيل الأوّل.

خفيف الثقيل: ما يساوي زمان دور واحد منها زمان الثقيل الأوّل.

ثقيل الرمل: ما يكون زمان كلّ دور من أدوارها[4] اثنى عشر سبباً[5] ثقالاً.

الرمل: ما يكون زمان كلّ دور منها اثنتى عشرة نقرة.

خفيف[6] الرمل: ما يكون زمان كلّ دور منها عشر نقرات.

الهزج: ما يساوي زمان[7] دور واحد منها زمان خفيف الرمل.

الفاختي: ما يكون زمان دور واحد منها عشرين نقرة.

الطنيني: أن يقسّم البعد ذو الأربع بمدّة ونصف مدّة، ثمّ نغمة المطلق، ثمّ السبّابة، ثمّ البنصر، ثمّ الخنصر.

القويّ والمقويّ والرجلي: كذلك.

اللوني:[8] أن يقسّم البعد ذو الأربع بنصف مدّة، ونصف مدّة وثلاثة أنصاف مدّة.

الملوّن والخنثويّ: كذلك.

التأليفي: أن يقسّم البعد ذو الأربع بربع مدّة ومدّتين.

النسوي[9] والناظم والراسم: كذلك.

والله أعلم.

[1] بينها: بينهما، آ. [2] أوضاع: اضاع، آ. [3] بالطباع: بالطبايع، آ س. [4] أدوارها: أدوار، س. [5] سبباً: -، آ س.
[6] خفيف: -، ب. [7] زمان: ~~زمان~~، ب. [8] اللوني: اللوي، ب. [9] النسوي: النسري، آ ب س.

الباب الثامن عشر: في النجوم

[**النجوم**:] وهي علم يعرف بها تأثيرات النجوم في السفليات مع دلائلها[1] ومنسوباتها.

وقيل: علم بقواعد يعرف بها أحكام تأثير النجوم وخواصّها.

الزيج: كتاب يحسب منه سير الكواكب ليستخرج منه التقويم.

الطالع: هو البرج الموافي[2] للأفق عند الولادة أو أمر ما.

الزايجة:[3] صورة مربّعة أو مدوّرة تعمل لمواقع الكواكب في الفلك لينظر فيها عند[4] الحكم لمولد الطفل أو غيره.

الهيلاج: عبارة عن دليل العمر.

الرابي:[5] هو المستولي على موضع الهيلاج، ويقال له الكذخداه.[6]

بيت الكوكب:[7] بيت ينسب إليه.

شرف الكوكب:[8] درجة في برج ينسب اليه.

الهبوط: مقابله.

الوبال: هو البرج المقابل لبيت الكوكب.

المثلّثة: كلّ ثلاثة أبرج تكون على طبيعة واحدة ينسب إليه ثلاثة كواكب.

الأوتاد: عبارة عن الطالع والرابع والسابع والعاشر.

ما يلي الأوتاد:[9] عبارة عن الثاني[10] والخامس والثامن والحادي عشر.

السواقط: عبارة عن الثالث والسادس والتاسع والثاني عشر.

حيّز الكوكب: كونه إذا كان ذكرياً نهارياً فوق الأرض بالنهار، وإذا كان ليلياً مؤنّثاً تحت الأرض بالليل.

[1] دلائلها: دلالتها، ب. [2] الموافي: الموفي، س. [3] الزايجة: الزابحة، ب. [4] عند: عدد، آ. [5] الرابي: كذا في جميع النسخ. ولم نجده في مصادر أخرى. [6] الكذخداه: الكذخداء، ب. [7] الكوكب: الكواكب، آ. [8] شرف الكوكب: تشرف الكوكب، ب؛ شرف الكواكب، آ. [9] الأوتاد: الأوتار، آ. [10] الثاني: + والخمس، آ.

 | DOI:10.1163/9789004423367_026

دستورية[1] الكوكب: كونه في بيته أو شرفه في الوتد، وكوكب آخر على هذه الصفة ناظر إليه.

المواجهة: كون الكوكب في بيته.

الاثنى عشرية: أن يؤخذ من أوّل البرج إلى الدرجة والدقيقة التي فيها الكوكب[2]، ويضرب في اثنى عشر، ويلقى ما اجتمع من أوّل البرج فحيث انتهى، ففي ذلك البرج وتلك الدرج اثنا عشرية الكوكب.

الأجرام: قدر الدرجات[3] التي بين أوّل الاتّصال.

الرباطات: هي الدرج التي بين الكواكب والشمس في حال رجوعه.

النقل: انصراف الكوكب عن كوكب واتّصاله بآخر.

المنع: أن يكون ثلاثة كواكب في برج واحد على درجة مختلفة.

دفع[4] الطبيعة: كون الكوكب في برج له فيه مزاعمة[5] واتّصاله بآخر له أيضاً في ذلك البرج.

وقيل: اتّصال كوكب نهاري بآخر نهاري، وهما في مكان النهار وبالعكس.

دفع[6] التدبير: هو الاتّصال إلى أيّ جهة كانت.

الانتكاث: رجوع الكوكب حالة ابتدائه باتّصال آخر.

الاعتراض: كون الكوكب ذاهباً إلى اتّصال كوكب وقبل البلوغ إليه يرجع[7] المتّصل به إلى برج آخر ويتّصل به كوكب آخر.

خلاء السير: انصراف الكوكب عن كوكب من غير أن يتّصل بآخر.

الوحشي: كون الكوكب في برج لا ينظر إليه فيه كوكب ولا ينظر هو إلى كوكب.

دفع القوّة: أن يكون الكوكب في شرفه أو أحد حظوظه ويتّصل بكوكب آخر، فيدفع قوّته.

الإقبال: كون الكوكب في الأوتاد وما يليها.

الإدبار: كونه في السواقط.

النغمة:[8] أن يكون الكوكب في هبوطه أو وباله فيتّصل به كوكب له مراغمة في ذلك البرج فيخرجه من هبوطه الكواكب.

المكافأة: أن يتّفق فيقع لذلك المتّصل أداة[9] حالة فيخرجه المتّصل به منه.

---

[1]دستورية: دوستورية، س. [2]الكوكب: -، آ س. [3]الدرجات: الدجات، آ. [4]دفع: رفع، آ ب س.
[5]مزاعمة: مراغمة، آ ب س. [6]دفع: رفع، آ س. [7]يرجع: رجع، آ. [8]النغمة: النعمة، آ س. [9]أداة: دواة، ب.

القبول: اتّصال الكوكب بكوكب من بيت المتّصل به أو شرفه أو مثلثه أو وجهه.

الردّ: اتّصال كوكب بكوكب منفرد لا يتّصل شيء من الكواكب.

السهم: دليل مستخرج من دليلين دالّين على شيء واحد.

القوّة الطبيعية: كون الكوكب في حظّ من حظوظه أو صاعداً في فلك أوجهة سريع السير مستقيماً مشرقاً أو صاعداً لعرض الشمال.

القوّة العرضية: أن يكون [الكوكب] في وتد أو ما يلي وتداً.

المراكز: مواضع القمر عند الاجتماع أو الاستقبال أو تربيعيته للشمس.

فتح الباب:[1] انصراف القمر عن كوكب واتّصاله بآخر بيته في مقابلة بيت الأوّل.

الآبار:[2] درج في البروج إذا بلغتها الكواكب نحست[3] فيها.

الابتزاز: أن يكون للكوكب[4] حظوظ كثيرة في البرج.[5]

الاستعلاء: أن يكون الكوكب في البرج[6] العاشر من العاشر[7] الآخر.

الحصار:[8] أن يكون مضغوطاً[9] بين نحسين.

الاحتراق: أن يكون مقارناً للشمس وبينهما أكثر من دقائق الصميم.

الفردار: قسمة للعمر[10] بين الكواكب السبعة.

التسيير: أن ينظر حكم[11] بين الهيلاج وبين السعد والنحس فيؤخذ لكلّ درجة سنة.

التربيع: أن يصير الكوكب من آخر على ربع الفلك.

التثليث: أن يصير منه على ثلث الفلك.

التسديس: أن يصير منه على سدس الفلك.

المقابلة: أن يصير منه على نصف الفلك.[12]

الاجتماع: يعبّر به عن المحاق.

القِران: يعنى به[13] اجتماع زحل والمشتري خاصّة.

والله أعلم.

---

[1]الباب: -، ب. [2]الآبار: الآباد، ب. [3]نحست: بحسب، ب. [4]للكوكب: للكواكب، ب. [5]البرج: البروج، آ س. [6]البرج: برج، آ؛ البيع، ب. [7]العاشر: -، آ ب. [8]الحصار: الحضار، ب. [9]مضغوطاً: معطوطاً، آ ب س. [10]للعمر: للقمر، آ س. [11]أن ينظر حكم: أن ينظركم، آ ب س. [12]الفلك: اللفلك، ب. [13]به: -، ب.

الباب التاسع عشر: في الطبّ

علم الطبّ: ما يعرف منه أحوال بدن الإنسان من جهة ما يصحّ ويزول عن الصحّة[1] لتحفظ الصحّة حاصلة وتستردّ زائلة.

الطبّ النظري: ما لا يتعرّض فيه لبيان كيفية عمل.[2]

الطبّ العملي:[3] ما يتعرّض فيه لذلك.[4]

الصحّة: هيئة بدنية يكون الأفعال بها لذاتها سليمة.[5]

المرض: هيئة مضادّة لها.

العرض: ما يتبع المرض.

السبب: ما يوجب الآفة في البدن بواسطة.

مرض التركيب: هو الواقع في الخلقة والعدد والمقدار.

مرض الشكل: أن يصير شكل العضو على صفة تضرّ[6] بالأفعال.

الرياضة: حركة إرادية تضطرّ إلى التنفّس العظيم المتواتر.

التدبير: التصرّف في الأسباب الضرورية.

الاستعداد: تقديم الدلك على الرياضة.

الاسترداد: تأخيره عنها.

النبض: حركة وضعية للشرايين قبضاً وبسطاً لتعديل الروح بالنسيم وإخراج فضلاته.

فصل

القوّة: مبدأ[7] كلّ فعل في البدن.

القوّة الحيوانية: ما تعدّ البدن لقبول الحياة وأفعالها.

---

[1] الصحة: صحة، ب. [2] عمل: العمل، آب. [3] العملي: العمل، ب. [4] لذلك: كذلك، آ. [5] سليمة: في ب عليها خطّ ترقين، وصحّحت في الهامش. [6] تضرّ: يضرّ، ب. [7] مبدأ: مبتدأ، ب.

 | DOI:10.1163/9789004423367_027

القوّة الطبيعية: ما يتصرّف في الغذاء لبقاء الشخص، وبعد تصرّفها لبقاء النوع.

الغاذية: ما تحيل الغذاء إلى مشابهة المغتذي.

النامية: ما تزيد[1] في أقطار البدن على نسبة مخصوصة حتّى تبلغ به كمالها[2] المقداري ثمّ يبطل فعلها.

المولّدة: ما تحصّل المنيّ ليستعدّ لقبول صورة بدن إنساني.

المصوّرة: ما تصدر عنه الأفعال المتعلّقة بنهايات مقادير البدن.

الجاذبة: ما تجذب المنافع إلى البدن.

الماسكة: ما تمسكها لتصرّف القوى المحتاج إلى تصرّفها.

الهاضمة: ما تحيل الغذاء إلى القوام الصالح بترقيق ما غلظ وتغليظ ما رقّ.

الدافعة: ما تميط عن البدن ما لا حاجة إليه.

فصل

الدواء: ما يؤثّر في البدن بكيفيته فقط.

الغذاء: ما يؤثّر فيه بمادّته فقط.

الدواء اللطيف: أن يصغر عند فعل حرارتنا فيه.[3]

الكثيف: يقابله.

المحلّل: ما يهيّئ المادّة للتبخير.

المحلّل للرياح: ما يرقّق[4] الريح ليندفع.

الملطّف:[5] ما يجعل المادّة أرقّ.

المغلّظ: ضدّه.

المكثّف: ما يستحصف منه مسامّ البدن.

اللَزِج: ما لا ينقطع[6] عند[7] الامتداد.

---

[1]تزيد: تزيل، ب. [2]كمالها: كمافها، ب. [3]أن يصغر عند فعل حرارتنا فيه: أن ينعصر عنه فلعل جرّار ينافيه، آ س؛ أن ينصعر عنه فعل جرّار ينافيه، ب. [4]يرقّق: يرقّ، آ. [5]الملطّف: المكلف، ب. [6]لا ينقطع: ينقطع، آ س. [7]عند: عن، آ س.

الهَشّ: ما يتفتّت بأدنى مسّ.

الجامد: ما من شأنه أن يسيل وهو في الحال مجتمع.

السائل: من شأنه أن يبسط أجزاؤه إلى أسفل.

اللعابي: ما ينفصل منه إذا نقع أجزاء ويصير المجموع لزجاً.

الدهني: ما في جوهره دهن.

المنشّف: ما إذا لاقته مائية غاضت في مسامّه[1]، فلا يظهر فيه أثر.

الجالي: ما يجرّد الرطوبة اللزجة عن مسامّ[2] العضو.

المُخَشّن: ما يجعل أجزاء سطح العضو مختلفة الوضع بعد ملاسة طبيعية أو عارضة عن مادّة لزجة.

المفتّح: ما يخرج المادّة السادّة عن المجرى إلى الخارج.[3]

المرخّي: ما يليّن العضو بحرارته ورطوبته.

المنضِج: ما يعدل قوام الخلط ويهيّئه للدفع.

الهاضم: ما يفيد الغذاء سرعة إنضاج.[4]

المفجّح: ضدّه.

المقطّع: ما يقسّم المادّة إلى أجزاء[5] صغار، وإن بقيت على غلظها.

الجاذب: ما يحرّك[6] المادّة إلى موضعه.

الرادع: ضدّه.

اللاذع:[7] ما يفرّق اتّصال العضو بقوّة نفّاذة في مواضع لا تحسّ بانفرادها[8] بل بجملتها.

المحمّر: ما يجذب الدم بقوّة إلى الجلد مع تسخين.

المقرّح: ما يفني الرطوبة الأصلية ويجذب مادّة رديّة.

المحكّك: ما يجذب خلطاً لذّاعاً[9] حارّاً.

المحرّق:[10] ما يفني لطيف الأخلاط ويبقى رماديّتها[11] لحرارته.

الأكّال:[12] ما يبلغ من تقريحه وتحليله أن ينقص[13] قدراً من اللحم.

---

[1] مسامه: مسافه، آ؛ مسامة، س. [2] مسام: مسلم، آ س. [3] الخارج: خارج، آ ب س. [4] إنضاج: ايضاج، ب. [5] أجزاء: آخر، آ. [6] يحرّك: يتحرّك، آ. [7] اللاذع: اللارع، س. [8] بانفرادها: بأفرادها، ب. [9] لذّاعاً: لدّاغاً، آ ب س. [10] المحرّق: لمحرّق، آ. [11] رماديّتها: رمادتها، آ ب س. [12] الأكّال: الأكل، آ. [13] أن ينقص: أن ينقض، آ.

المفتّت: ما يصغّر أجزاء الخلط المتحجّر.

المعفّن: ما يفسد مزاج الروح والرطوبة الأصلية حتّى لا تصلح لما أعدّت له.

الكاوي: ما يأكل اللحم ويحرق[1] الجلد ويجفّفه ويصلّبه كالحممة.

القاشر:[2] ما يبلغ من جلائه[3] إلى إخراج الأجزاء الفاسدة.

المقوّي: ما يعدّل مزاج العضو وقوامه حتّى لا يقبل الفضول.

المخدّر: ما يجعل الروح الحسّاس والمحرّك[4] للعضو غير قابل للتأثير النفساني قبولاً تامّاً.

المنفّخ: ما فيه رطوبة فضلية لا تقوى الحرارة على تحليلها[5] بل يستحيل رياحاً.

الغسّال: ما يجلو[6] المادّة برطوبته وسيلانه لا بجلائه.[7]

الموسّخ: ما يرخي القروح برطوبته.

المزلّق: ما يبلّ سطح الفضلة المحتبسة في المجرى.

المملّس: ما يبسط على سطح عضو خشن فيستر[8] خشونته.

المجفّف: ما يفني الرطوبة بتلطيفه وتحليله.

القابض: ما يجمع[9] أجزاء [العضو].[10]

العاصر: ما يبلغ قبضه إلى إخراج ما في تجويف العضو.

الدافع: ما يزيل الموادّ من الظاهر إلى الباطن بالدفع القويّ.

المسدّد:[11] ما يحتبس في المجرى[12] لكثافته أو يبوسته أو عروشه.[13]

المغري: يابس ذو رطوبة لزجة يلتصق على الفوّهات فيسدّها.

المدمّل: مجفّف[14] يجعل الرطوبة التي بين شفتي الجرح لزجة فتلتصقان.

المنبت[15] **اللحم**: ما يعقد الدم الوارد إلى الجراحة لحماً.

الخاتم: ما يجعل على سطح الجراحة خشكريشة[16] تكنّها[17] من الآفات.

---

[1] يحرق: يخرق، آ س. [2] القاشر: الفاشر، آ ب س. [3] جلائه: خلائه، آ س. [4] المحرّك: المحترک، آ. [5] تحليلها: تحليها، س. [6] يجلو: ينجى (؟)، آ س. والصواب "يجلو" كما في الصيدلية المجرّبة لابن نفيس والقانون لابن سينا. [7] بجلائه: بجالائه، آ س. [8] فيستر: فينشر، آ س؛ فينتشر، ب. [9] يجمع: يجتمع، آ ب س. [10] [العضو]: كما في الصيدلية المجرّبة. [11] المسدّد: المسرد، ب. [12] المجرى: المخبري، آ س. [13] عروشه: كذا في آ و س؛ عروسة، ب. [14] مجفّف: مخفف، ب س. [15] المنبت: المنبث، آ. [16] خشكريشة: عشكريشة، آ س. [17] تكنّها: يلها، آ س؛ تليها، ب.

الترياق: كلّ ما يحفظ صحّة الروح وقوّته ليتمكّن من دفع السموم.

الفاذزهر: مثله.

فصل

الأعضاء: أجزاء متماسكة يتألّف منها البدن.

العضو المفرد: ما لكلّ جزء محسوس منه اسم الكلّ[1] وحدّه.[2]

المركّب: يقابله.

الأعضاء الرئيسة:[3] ما بها بقاء الشخص أو النوع.

العصب: أجسام تنبت عن الدماغ أو النخاع، بيض، لدنة[4] في الانعطاف، صلبة في الانفصال.

العضلة: عضو مركّب من العصب ومن جسم ينبت من أطراف العظام شبيه[5] بالعصب.

الأوتار: أجسام تنبت من أطراف بعض العضل شبيهة بالعصب.

الرباطات: عصبانية المرأى والملمس تنبت في الأعضاء[6] إلى مؤخّر العضل.

الشريان: جسم ينبت من القلب، مجوّف، رباطي الجوهر[7]، يتحرّك بحركات انبساطية وانقباضية للترويح ونفض البخار[8] الدخاني.[9]

الوريد: عرق شبيه بالشريان إلّا أنّه نابت من الكبد.[10]

الغشاء: جسم منتسج[11] من ليف عصباني غير محسوس قليل الثخن.

الباسليق:[12] عرق في اليد[13] عند[14] المرفق في الجانب الإنسي الأيسر إلى ما يلي[15] الإبط.

القيفال: هو عند المرفق في الجانب الوحشي.

الأكحل: بين الباسليق[16] والقيفال.

---

[1] الكلّ: لكلّ، آ س. [2] وحدّه: واحد، ب. [3] الرئيسة: الرقية، س. [4] لدنة: لذة، آ س. [5] شبيه: شبيهة، آ س. [6] تنبت في الأعضاء: تنقب من الأعضاب، آ؛ تنقب من الأعضاء، ب س. [7] الجوهر: الجوهري، ب. [8] البخار: البخاري، س. [9] الدخاني: -، آ. [10] في الكبد: بالكبد، آ س. [11] منتسج: منبتيح، آ. [12] الباسليق: الباسيلق، آ. [13] اليد: البدن، آ. [14] عند: عنه، ب. [15] الإنسي الأيسر إلى ما يلي: الانسبى الأيسر مائل، آ س. [16] الباسليق: الباسيليق، آ س.

الودجان:[1] العرقان في العنق من جانبي الحلقوم.

حبل الذراع: في ظاهر الساعد[2] من شعب القيفال.

الأسيلم: عرق بين الخنصر والبنصر.

الصافن: عرق في الساقين يظهر عند الكعب الداخل في الجانب الإنسي.

عرق النَّسا: قبالة الصافن من الجانب الوحشي.

البوّاب: معىً متّصلة بالمعدة من أسفل ينضمّ عند دخول الطعام[3] إلى المعدة إلى أن ينهضم ثمّ ينفتح.

الاثنا عشر:[4] معىً يتّصل بالبوّاب.[5]

الصائم: معىً يلي الاثنا عشر[6]، يلبث فيها الطعام.

القولون: المعى الذي يحدث فيه القولنج.

الأعور: معىً على هيئة الكيس لا منفذ لها.

الحجاب:[7] عضو شبيه بالجلد يأخذ من رأس القصّ إلى الظهر فيتّصل بتجويف البطن.

مراقّ البطن: ما رقّ منه.

فصل

الحمّى: حرارة غريبة ضارّة[8] بالأفعال، تنبعث من القلب إلى الأعضاء.

الورد: الحمّى النائبة في كلّ يوم.

الغبّ: الحمّى التي تنوب يوماً ويوماً [لا].

الربع: التي تنوب يوماً ويومين [لا] ثمّ تعود[9] في الرابع.

الخمس والسدس: على هذا القياس.

المطبقة: الدائمة التي لا تنقطع.[10]

سونوخس: حمّى تحدث عن غليان الدم.[11]

---

[1] الودجان: الوجدن، آ، الوجدان، س. [2] الساعد: الساعة، ب. [3] الطعام: العطام، ب. [4] الاثنا عشر: الاثنا عشري، ب س. [5] البوّاب: البوابه، س. [6] الاثنا عشر: الاثنا عشري، آب س. [7] الحجاب: المحاب، آس. [8] ضارة: صادة، آس. [9] تعود: تنوب، آس. [10] لا تنقطع: تنقطع، آس. [11] الدم: دم، آب س.

البحران: تغيّر عظيم يحدث دفعة إلى الصحّة أو العطب.

الصداع: ألم في أعضاء الرأس.

البيضة: صداع مزمن يهيّج كلّ ساعة مع كراهة الضوء والكلام.

الخوذَة: مثلها.

الشقيقة: بيضة تخصّ شقّاً من الرأس.

السَرْسام: ورم حارّ عن صفراء، أو دم صفراوي في أحد جانبي الدماغ الداخلين.

السبات: ورم دماغي عن بلغم وصفراء.

الحمق: نقصان في المفكّرة وبطلان عن برد[1] ساذج أو مادّي أو يبس أو هما معاً.

النسيان: نقصان أو بطلان لقوّة[2] الذاكرة.[3]

مانيا: جنون سبعي[4] عن سوداء محترقة عن دم أو صفراء أو داء[5] يكون مع اضطراب وتوثّب.

ماليخوليا: تشويش الظنون والفكر إلى الفساد والخوف.

داء[6] الكلب: نوع من المانيا مع معاشرة وموافقة وكثير[7] ضحك.

الوباء: فساد يعرض لجوهر الهواء لأسباب سماوية أو أرضية.

الشخوص: أن يكون ملقىً لا يطرف وهو شاخص.

السهر: يقظة مفرطة[8] عن يبس وحرّ[9] يحدّان الروح ويوجبان حركتهما إلى خارج.

الدوار:[10] ظلمة تعتري البصر عند القيام.

وقيل: هو أن يتخيّل الأشياء تدور.

السدر:[11] مقدّمة الدوار.

الكابوس: أن يتخيّل في النوم خيالاً يقع عليه ويعصره بضيق النفس ومنع الحركة.

ليثرغس:[12] ورم عن بلغم عفن في مجاري روح الدماغ.

الصرع: سدّة دماغية غير تامّة يتشنّج[13] بها الأعصاب لانقباض مبدأها مع امتناع الحسّ والحركة والانتصاب.

---

[1] برد: ورد، آس. [2] لقوّة: القوّة، آ. [3] الذاكرة: الذكرة، آ؛ الذكر، ب س. [4] سبعي: سبقى، آ. [5] داء: داع، آس. [6] داء: داع، آس. [7] كثير: قليل، ب. [8] مفرطة: مفرط، آس. [9] حرّ: + ~~عن~~، س. [10] الدوار: ~~الدوار~~ الدوار، آ. [11] السدر: السدر~~س~~، س. [12] ليثرغس: ليسترغس، آس. [13] تتشنّج: تشنّج، آس.

السكتة: سدّة تامّة في بطون الدماغ ومجاري روح تعطّل الأعضاء عن الحسّ والحركة إلّا التنفّس[1] لضرورة الاستنشاق.

الفالج: استرخاء شقّ البدن طولاً.

التمدّد: مرض آلي يمنع انقباض الأعصاب.

اللقوة: مرض آلي ينجذب له شقّ الوجه إلى جهة غير طبيعية.

الرعشة: مرض يحدث عن عجز القوّة المحركة عن تحريك العضل[2]، أو ثباته على الاتّصال فيختلط حركات إرادية أو ثبات إرادي بحركة نقل العضو إلى أسفل.

الخدر: علّة آلية تحدث في الحسّ[3] اللمسي نقصاناً.

اليرقان: تغيّر فاحش في اللون[4] إلى صفرة أو سواد أو اجتماعهما.

فصل

التكدُّر: تسخّن وترطّب يعرض للعين يشبه الرمد.

الرَمَد: ورم حارّ في الملتحمة عن مادّة في العين، أو منحدرة[5] عن الرأس.

السَبَل: غشاء معروق أحمر غليظ[6] يحدث على البياض والسواد.

الوَردينَج: رمد عظيم يرم فيه البياض حتّى يمنع التغميض.

الظفرة: زيادة في الملتحمة أو الغشاء المجلّل للعين يبتدئ من الموق[7] الإنسي في الأكبر.

الجساء:[8] عسر فتح العين عند الانتباه من النوم.

القمر:[9] فساد يصيبها[10] إذا نظرت كثيراً إلى الثلج.

السلاق: غلظ في الأجفان عن مادّة غليظة رديئة أكّالة.

البردة: رطوبة تتغلّظ[11] وتتحجّر في باطن الجفن.

---

[1] التنفّس: النفس، آ. [2] العضل: الفضل، آ س. [3] في الحسّ: -، آ س. [4] في اللون: من الكون، آ؛ من اللون، س. [5] منحدرة: متحرزة، آ؛ متحذرة، س. [6] معروق أحمر غليظ: معروف حمر غلاظ، آ س؛ معروق حمر غلاظ، ب. [7] الموق: المرق، س. [8] الجساء: الجساد، آ. [9] القمر: القمس، آ س. [10] يصيبها: يصلبها، آ س. [11] تتغلّظ: تتلغط، آ.

الشعيرة:[1] ورم مستطيل يظهر على طرف الجفن.

السرتاق: زيادة شحم في الجفن الأعلى يثقله ويجعله كالمسترخي.

الخيالات: أشكال ذات ألوان[2] ترى في الجوّ.

الجَرَب: خشونة تحدث داخل الأجفان فتؤذي الحدقة.

الدَمْعَة: أن تكون آماق العين أبداً رطبة من غير رمد ولا وجع.

الطرفة:[3] أن تحدث فيها نقطة حمراء من ضربة أو غيرها.

الانتشار: أن يتّسع ثقب الناظر حتّى يلحق البياض من كلّ جانبيه.

الغرب: أن يترشّح[4] ماء في العين ويسيل منه متى غمز صديد.

الناصور: هذا، وقد يكون في موضع آخر.

الماء:[5] رطوبة غريبة تحتبس في ثقب العني بين الصفاق والرطوبة البيضية.

الماشرا: ورم حارّ عن دم صفراوي يعمّ الوجه.

البادوشنام:[6] حُمرة مفرطة تعرض في الوجه شبيهة بابتداء الجذام.

السعفة: قروح يابسة في الرأس أو الوجه يسيل منها الصديد.

الإ برية والهبرية: كالنخّالة فيه.

الشرى: ما يأخذ من الجلد أحمر كهيئة الدراهم.

الخنّاق: امتناع النفس أو البلع أو تعسّرهما لمزاحمة أو عجز قوّة أو ورم.

الربو: هو عسر النفس.

نفس الانتصاب: أن لا يتأتّى[7] النفس إلّا بانتصاب الرقبة ومدّها إلى فوق.

البواسير في الأنف: أن ينبت لحم داخل الأنف فيحتشي[8] به.

الخَشَم: فقدان حاسّة الشمّ.

القلاع: بثور في الحنك واللسان.

الضفدع: غدّة[9] تنعقد تحت اللسان.

---

[1]الشعيرة: الشعير، آ س. [2]ألوان: اللوان، آ. [3]الطرفة: الظرفة، آ س. [4]أن يترشّح: يترشّح، ب. [5]الماء: المساء، آ. [6]البادوشنام: البازدشنام، آ س. [7]أن لا تأتّى: أن لا يأتي، آ، أزلاً ياتى، س. [8]فيحتشي: فتحشى، ب. [9]غدّة: عقدة، ب.

السلعة: زيادة تحدث في الجسد[1] تتحرّك إذا حركت قدره من حمّصة[2] إلى بطيخة.

الحفر في الأسنان: ما يلزق بها من باطن أو[3] ظاهر.

فصل

الحصف: بثور[4] تهيج من كثرة العرق.

النملة: بثور[5] صغار تحدث[6] عن[7] صفراء حرّيفة لطيفة.

الجمرة: بثر أكّال منفط[8] محرق محدث للخشكريشة.

النار الفارسية: نفّاطات ممتلئة ماءً رقيقاً تخرج بعد حكة ولهب.[9]

الثآليل: نكت تخرج[10] باليد يابسة لا ترشح كالعدسة.

الفلغموني: ورم حارّ يضرب في لونه حمرة.

الداخس: ورم يظهر عند الأظفار شديد الضربان.

السِلّ:[11] قرحة في الرئة يلزمها حمّى دقيقة للقرب من القلب ونفث المدّة.

الخفقان: اختلاج يعرض للقلب ليدفع به المؤذي.

الغُشي: حالة يتعطّل معها[12] الحسّ والحركة لضعف القلب.

الفواق: حركة فم المعدة لدفع ما يؤذيه.

القولنج: وجع معويّ يعسر معه خروج ما يخرج بالطبع.

الشوصة: ريح ينعقد في الأضلاع.

الاستسقاء: مرض ذو مادّة غريبة يتخلّل الأعضاء فيربوها.[13]

الزقّي:[14] أن ينتفخ البطن وتنتؤ السرة[15] ويسمع خضخضة إذا حرّك.

الطَبْلي: أن يكون البطن منتفخاً متمدّداً يسمع منه صوت الطبل.

---

[1] الجسد: الجيد، آ ب س. [2] حمّصة: خمصة، س. [3] أو: و، آ س. [4] بثور: ثبور، س. [5] بثور: ثبور، س. [6] صغار تحدث: ضعار تحث، آ. [7] عن: + ضوء، آ. [8] منفط: متفط، آ؛ متقط، س. [9] لهب: لهيب، آ ب س. [10] تخرج: -، آ. [11] السل: الستل، آ س. [12] معها: منها، آ س. [13] فيربوها: فتربوها، آ س. [14] الزقّي: الذقي، س. [15] تنتؤ السرة: يلتو السدة، آ؛ يليو السرّة، ب؛ يلق السدة، س.

اللحمي: أن يكون في الجفن[1] والأطراف ورم رخو ترمّ الأسباب[2] وترهّل[3] الوجه واليدين.

ذات الجنب: وجع تحت الأضلاع ناخس مع سعال وحمّى.

السلّ: أن يتناقص اللحم بعد سعال مزمن[4] وتعب شديد.

الدِقّ:[5] حمّى تدوم ولا تدوم، لا تكون قوية، تفضي[6] إلى ذبول وضني.

الشهوة الكلبية: دوام الجوع مع الأكل الكثير فمتى أكل يثقل عليه فيقيئه أو يغثيه.[7]

الخلفة: أن لا يلبث الطعام في البطن اللبث المعتاد.

الرجاء:[8] علّة تحدث للمرأة تشبه حالها حال الحبلى في عظم البطن وفساد اللون واحتباس الطمث.

الجساءة:[9] صلابة في الكبد عن ورم يحدث، ولا يتحلّل، ويصلب، وينعقد.

النِقْرِس: ورم المفاصل لموادّ تنصبّ إليها.

عرق النساء: وجع يمتدّ[10] من الورك إلى الفخذ كلّه في مكان منه بالطول، وربّما بلغ الساق والقدم ممتدّاً.

الفَتْق:[11] أن يكون بالرجل نتوء في مراقّ البطن، فإذا استلقى وغمزه إلى داخل غاب، وإذا استوى عاد.

القرو: أن يعظم جلد البيضتين لريح أو ماء أو لنزول الأمعاء أو الثرب.[12]

ديابيطس:[13] أن يدوم العطش، وكلّما شرب بال.

العذيوط: أن يكون كثير الشبق، رخو[14] المقعد.[15]

الدوالي: عروق تظهر في الساق غلاظ ملتوية[16] شديدة الخضرة والغلظ.

داء الفيل: أن يتورّم الساق كلّه ويغلظ.

العرق المدني: ما يخرج عن الساق بعد لهيب يحدث[17] في ذلك الموضع فيلفّ ما يخرج منه على قصبة فينجرّ قليلاً قليلاً.

---

[1] الجفن: الحقو، آ ب س. وفي مفاتيح العلوم للخوارزمي: الأجفان. [2] ترمّ الأسباب: وفي مفاتيح العلوم: وترمّ الأنثيان. [3] وترهّل: وفي مفاتيح العلوم: يترهّل. [4] مزمن: مذمن، آ؛ منه من. [5] الدقّ: الدو، آ س. [6] تفضي: تقضي، ب. [7] فيقنيه أو يغثيه: فتقيئه أو يغبثه، آ. [8] الرجاء: الرخاء، آ س. [9] الجساءة: الجساوة، آ ب س. [10] يمتدّ: تميد، آ. [11] الفتق: الفتولن، آ؛ الفتو، س. [12] أو الثرب: والثرب، آ س؛ والأثرب، ب. [13] ديابيطس: ديانبطس، آ؛ ديانيطس، س ب. [14] رخو: زحو، آ س؛ لعو، ب. [15] المقعد: المعقد، ب. [16] ملتوية: ملتومة، آ س. [17] يحدث: الحدث، آ س.

الباب العشرون: في الأخلاق

علم الأخلاق: هو العلم بالأحوال التي تخصّ شخصاً واحداً.

الخلق: ملكة نفسانية تصدر عنها[1] الأفعال[2] النفسانية بسهولة من غير رويّة.

وقيل: هو اسم جامع للقوى المدركة بالبصيرة. وتجعل تارة للقوى الغريزية، وتارة للقوى المدركة[3] التي يصير الإنسان بها[4] خليقاً أن يفعل شيئاً دون شيء.

قوّة النزوع: ما بها[5] يكون الطلب للموافق والهرب من المخالف والرضا والغضب والإيثار والكرامة.[6]

قوّة التفكّر: ما بها يكون النطق والعقل والعلم[7] والحكمة والرويّة والمهنة والرأي والمشورة.

قوّة الغذاء: ما بها النشؤ والتربية والولادة.

قوّة الحسّ: ما بها يكون الإحساس واللذّة والألم.

الخلق: اسم جامع للهيئات والأشكال والصور المدرَكة[8] بالبصر.

الطبع: حصول صور مخصوصة لا سبيل إلى تغييرها.[9]

الطبيعة: كذلك، اعتباراً بطبع السيف.

الضريبة: هكذا، اعتباراً بضرب الدرهم.

الغريزة:[10] لما غرز عليه.

النحيتة: كذلك، اعتباراً بالنحت والنحز.

الشيمة: الحالة التي عليها الغريزة اعتباراً بالشامّة اللتي في أصل الخلقة.

السجية: ما سجي عليه الانسان.

العادّة: تكرير الفعل والانفعال.

العقل: هو القوّة المتهيّئة لقبول العلم.

النهى: العقل المنتهي من المحسوسات إلى معرفة ما فيها من المعقولات.

---

[1] عنها: عنه، س. [2] شخصاً واحداً [...] عنها الأفعال: -، آ. [3] للقوى المدركة: للحالة المكتسبة، ب. بعد "للحالة" في هذه النسخة ورقة كتب عليها "بياض صح". [4] يصير الإنسان بها: بها يصير الإنسان، ب. [5] ما بها: بانها، آ س.
[6] الكرامة: الكراهة، ب. [7] والعلم: والحلم، آ؛ العلم، ب. [8] المدركة: المدرک، ب. [9] تغييرها: تعسرها، ب.
[10] الغريزة: العرندة، ب.

 | DOI:10.1163/9789004423367_028

الحجر: ما يلتزمه الإنسان من حظر[1] الشرع والدخول في أحكامه.

الحجا: ما يكون الإنسان به قاطعاً عما يقبح.

اللبّ: ما يخلص به من عوارض الشبه ويترشّح[2] لاستفادة[3] الحقائق من دون المفزع إلى الحواسّ.

العلم: إدراك الشيء بحقيقته. ويطلق على حصول صور المعلومات في النفس.

وقيل: حكم النفس على الشيء بوجود شيء له هو موجود أو نفي شيء عنه هو غير[4] موجود له.

الدراية: المعرفة المدركة بضرب[5] من الحيل، وهو تقديم المقدّمة وإجالة الخاطر واستعمال الرويّة.

الفضيلة: ما يحصّل به الإنسان[6] مزيّة على غيره متوصّلاً به إلى السعادة.

النافع: ما يعين على بلوغ الفضيلة والسعادة والخير.

الحسب: ما يختصّ به الإنسان فيعدّه من مآثره.[7]

الشرف:[8] ما[9] يؤثر عن الآباء.

الجمال: الهيئة التي لا تنبو[10] الطباع السليمة عن النظر اليها.

الدولة: سعادة دنيوية لا يعتبر فيها العاقبة.[11]

الاتّفاق:[12] مطاوعته التوفيق في السعادة والشقاوة.

التوفيق: موافقة إرادة[13] الإنسان وفعله قضاء الله تعالى وقدره.

التأييد: تقوية أمره من داخل بالبصر ومن خارج بالقوّة والبطش.

الرشد: عناية إلهية تعين الإنسان في الباطن عند توجّهه في أمور[14] فتقويه على ما فيه صلاحه[15] وتغيّره عما فيه فساده.

العصمة: فيض إلهي يقوى بها الإنسان على تحرّي الخير[16] والتجنّب عن الشرّ.

القصور: الخلوّ عن المعاني الباعثة على تحرّي الخيرات والفضائل وعدم التمكّن من أسبابها.

الخاصّ: من تخصّص من المعارف بالتحقيقات دون التقليدات.

العامّي: من يرضي[17] من المعارف بالتقليدات.

---

[1] حظر: حفظ آس؛ خط، ب. [2] يترشّح: هرسبح، آب. [3] لاستفادة: الاستعادة، ب؛ لاستعارة، آس. [4] غير: غيره، آس. [5] بضرب: يضرب، س. [6] الإنسان: للإنسان، ب. [7] يعدّه من مآثره: يعدّه مأثر، آ؛ يعدّه مأثره، س. [8] الشرف: الشوق، آ. [9] ما: -، ب. [10] لا تنبو: لا بهوا، س. [11] العاقبة: العافية، آس. [12] الاتّفاق: الإنفاق، آس. [13] إرادة: اراد، آس. [14] أموره: أمور، ب. [15] صلاحه: حلاحه، ب. [16] الخير: الخيرات، آس. [17] يرضي: مرضى، ب.

وقيل: الخاصّ من تخصّص في أمور البلد[1] بما يخرم بافتقاده[2] إحدى السياسات المدنية، والعامّي من لا يخرم بافتقاده شيء منها.

وقيل: الخاصّ من يسوس ولا يساس، والعامّي من يساس ولا يسوس، والوسط من يسوسه من فوقه ويسوس[3] هو من دونه.

فصل

الحكمة: اعتدال القوّة النطقية.

الذهن: جودة الفهم في إدراك ما وقع فيه التنازع.

الفطنة: سرعة ما يقصد إشكاله.

صفاء الذهن: استعداد النفس لاستخراج المطلوب بلا تشويش.

الفهم: ما يميّز ويدبّر كلّ فعل أنّه عدل أو ظلم.

وقيل: إدراك الأشياء الجزئية.

جودة الفهم: صحّة الانتقال من الملزومات إلى اللوازم.

الذكاء: سرعة اقتراح[4] النتائج.[5]

وقيل: المضاء[6] في الأمر وسرعة القطع بالحقّ.

حسن التصوّر: البحث عن الأشياء بقدر ما هي عليه.

سهولة التعلّم: قوّة النفس على إدراك المطلوب بلا زيادة سعي.

الحفظ: ضبط الصور المدركة.

وقيل: المواظبة على مراعاة الشيء وقلّة الغفلة عنه.[7]

الذكر: استحضار المحفوظات.

البديهية: معرفة تامّة تجيء بلا فكر ولا قصد.

الرويّة: ما كان من المعرفة بعد فكر كثير.

---

[1] في أمور البلد: من أمور من البلد، آ س. [2] بافتقاده: بافتقاره، آ؛ بانتقاده، س. [3] يسوس: يسو، ب.

[4] اقتراح: اقتداح، آ ب س. [5] النتائج: التنامح، آ. [6] المضاء: المضاد، آ. [7] عن: -، آ س.

الكيس: القدرة على جودة استنباط ما هو أصلح في بلوغ الخبر.

الخبر: المعرفة المتوصّل إليها.

الظنّ: إصابة المطلوب بضرب من الأمارة.

النظر: إجالة الخاطر[1] نحو المرئي لإدراك البصيرة إيّاه.

الرأي: إجالة الخاطر في رؤية ما يريده.

الخاطر: حركة الفهم نحو الشيء في ما لا يقبله النفس.

الوهم: انقياد النفس لقبول أثر ما يرد[2] عليها فيما يقبله النفس.

الخيال: انقيادها لذاك[3] القبول اعتباراً بما يكون من جهة الحاسّة.

التخيّل: تصوّر أعيان الأشياء بعد غيبوبتها عن الحسّ.

التفكّر: جولان القوّة المفكّرة بين الخواطر بحسب نظر العقل.

الذكر: وجود الشيء في القلب أو في اللسان.

الجربزة: استعمال الدهاء في الأمور الدنيوية مع الاقتصار بالأرباح[4] الدنية.[5]

الخبّ: استعماله[6] في تلك الأمور صغيرها وكبيرها.

الغباوة: ترك النظر مع بعض العمل.

الغمارة: قلّة التجربة في الأمور العملية مع تخيّل سليم.

البله: قلّة التنبّه[7] على الأمور الدنيوية.

الحمق: نقصان العقل العملي.

وقيل: هو أن يكون[8] غرضه في نفس الأمر صحيحاً وسلوكه إليه خطأ.

الخرق: الجهل بالعلوم العملية بأن يفعل أكثر مما يجب أو أقلّ أو على[9] غير النظام المحمود.

الرقاعة:[10] أن يلصق بقلبه كلّ محال.

الرعونة: الإتيان بما يخرج عن الصواب.

الضلال:[11] اعتقاد الباطل حقّاً أو الكذب صدقاً أو القبيح جميلاً وبالعكس.

الوقاحة: لجاج النفس في تعاطي القبيح.

---

[1] الخاطر: الخواطر، آ س. [2] يرد: + وما، آ ب س. [3] لذلك: كذلک، آ. [4] بالأرباح: بالأرماج، آ س.

[5] الدنية: الدينية، آ س. [6] استعماله: استعمال، آ. [7] التنبّه: التنبيه، آ س. [8] أن يكون: أن لا يكون، س.

[9] على: -، آ س. [10] الرقاعة: الرعا قاعة، س. [11] الضلال: الضلالة، آ.

فصل

الشجاعة: اعتدال القوّة الغضبية وهي عبارة عن صرامة القلب في الأهوال وربط الجأش في المخاوف.

الغضب: ثوران دم القلب إرادة للانتقام.

النجدة:[1] عدم الجزع عند المخاوف.

الصبر: قوّة مقاومة الآلام والأهوال.

كبر النفس: استحقار الفقر والكبر والصغار.

عظم الهمّة: عدم المبالاة بسعادة الدنيا وشقاوتها.

الشهامة: الحرص على ما يوجب الذكر الجميل من العظائم.

الاحتمال: إتعاب[2] البدن في الحسنات.

الرقّة: التأدّب[3] عن أذىً يلحق الغير.

الحلم: الطمأنينة عند سورة الغضب.

التحلّم: إمساك النفس عن قضاء وطرها إذا هاجت.[4]

العفو: ترك المؤاخذة بالذنب.

الصفح: ترك التثريب عليه.

السكون: التأنّي في الخصومات والحروب.

التواضع: استعظام ذوي الفضائل ومن دونه في المال والجاه.

وقيل: الرضا بمنزلة دون ما يستحقّه فضله ومنزلته.

الخشوع: يقال باعتبار أفعال[5] الجوارح.

العزّة: التأبّي عن حمل المذلّة.

وقيل: الترفّع عما يلحقه غضاضة.

الحمية: محافظة الحرم والدين عن التهمة.

الغيرة: ثوران الغضب حماية عن الحرم.[6]

الحفيظة: الغضب المقتضي للغيرة.

---

[1] النجدة: النجد، ب. [2] الاحتمال إتعاب: لاحتمال اتغاب، آ؛ الاحتمال اتغاب، س. [3] التأدّب: التاذى، آس؛ التاذّب، ب. [4] هاجت: هاحب، ب. [5] أفعال: + المدح، س. [6] الحرم: الحرام، س.

التهوّر: الثبات المذموم في الأمور المعطبة.

الجنون: عارض يغمر العقل.

الجبن: الفزع المذموم من الأمور المعطبة.[1]

الفزع: ما يعتري[2] من الشيء المخوف.

الجزع: ما يعتري من الشيء المؤلم.

وقيل: تحرّك دم القلب عن من[3] فوقه مع ظنّ أن لا سبيل إلى الانتقام منه.

الوتر: تردّد دم القلب بين انقباض وانبساط على من يشكّ أنّه هل يقدر على الانتقام منه أو لا.

الحقد: يرادفه.

الحَرَد:[4] غضب يكون مع[5] قصد المغضوب عليه.

الفرق: ما يعتري[6] من شيء يضرّ.

الذعر: يرادفه.

الإشفاق: ما يعتري[7] من فوات محبوب.

الخوف: توقّع المكروه عن أمارة.

الخشية: خوف يشوبه تعظيم المخشي مع المعرفة به.

الوجل: استشعار عن خاطر ليس معه أمارة.

الرهبة: خوف مع تحرّز واضطراب.

الهيبة: رهبة جالبة للخضوع عن استشعار عظيم.

الرعب: امتلاء الجوف[8] عن الخوف.[9]

الكبر: رفع[10] الإنسان فوق قدره.

وقيل: ظنّه[11] بنفسه أنّه أكبر من غيره.

التكبّر: إظهاره لذلك.

الصلف: يقال اعتباراً لميل في عنقه.[12]

الصعر: لميل في خدّه.

---

[1] الجنون [...] المعطبة: -، آ. [2] يعتري: يعثرى، آ. [3] من: -، آ س. [4] الحرد: الحرو، آ س. [5] مع: معه، ب. [6] يعتري: يعثرى، آ. [7] الإشفاق ما يعتري: الانشقاق ما يعثرى، آ؛ الانشفاق ما يعترى، س. [8] الجوف: الخوف، آب س. [9] عن الخوف: -، ب. [10] رفع: دفع، ب. [11] ظنّه: + ~~أن لا~~، س. [12] عنقه: عفته، آ.

البأو: استعفاء[1] النفس بالرفع عن الانقياد للواجب.

الخيلاء:[2] أن يظنّ بنفسه[3] ما ليس فيها.

الزهو:[4] الاستطارة مع[5] الفرح بنفسه.

الفخر: المباهاة[6] بالفضائل الخارجية عنه.

العجب: أن يظنّ بنفسه استحقاق منزلة هو غير مستحقّ لها.

فصل

العفّة: اعتدال الشهوة وهي عبارة عن التحرّز عن تناول المشتهيات المخالفة للشرع والمروءة.

الشهوة: انبعاث النفس لنيل ما تتشوّقه.

الحياء: انحصار النفس عن خوف ارتكاب القبائح أو عن خوف ما يعاب.[7]

وقيل: انقباض النفس عن القبائح.

الصبر: حبس النفس عن متابعة الهوى.

الدعة: السكون عند هيجان الشهوة.

النزاهة: اكتساب المال من غير مهانة ولا ظلم، وإنفاقه في المصارف الحميدة.

القناعة: الرضا بما دون الكفاية.

الزهد: الاقتصار على القليل.

الورع: ملازمة الأعمال الجميلة.

وقيل: ترك التسرّع إلى تناول الأغراض الدنيوية.

وقيل: هو الإحجام عن المحارم.

وقيل: الوقوف عن الشبهات.

وقيل: الكفّ عن كثير من المباحات.

الظلف: الامتناع عن تناول الشهوات الدنية.

---

[1] استعفاء: استعصار، آ؛ استعصاء، س. [2] الخيلاء: الحيلان، آ. [3] بنفسه: نفسه، ب. [4] الزهو: الذهو، ب.
[5] مع: من، ب. [6] المباهاة: المباحات، آ. [7] يعاب: يعتاب، آ.

الوقار: التأنّي في التوجّه نحو المطالب.

الرفق: حسن الانقياد لما يؤدّي إلى الجميل.

حسن السمت: محبّة[1] ما يكمّل النفس.

الصدق: تحرّي الصواب في القول والعمل.

الانتظام: تقدير الأمور وتدبيرها بحسب المصالح.

السخاء: إعطاء ما ينبغي لمن ينبغي.[2]

الكرم: الإعطاء بالسهولة وطيب النفس.

وقيل: هو اسم لجماعة[3] الأخلاق المحمودة والأفعال المرضية إذا ظهرت بالفعل.

الإيثار: الإعطاء مع الكفّ عن حاجاته.

البذل: أن يكون مسروراً بإعطائه.

المواساة: أن يكون مع مشاركة الأصدقاء.[4]

السماحة: بذل مالم يجب تفضّلاً.

المسامحة: ترك ما يجب تنزّهاً.

الشكر: تصوّر[5] النعمة من المنعم وإظهارها باللسان والجوارح.

وقيل: هو الامتلاء من ذكر المنعم.

الفجور: عدم التمالك في تناول الشهوات.

الخمود: ضعف القوّة الشهوية عن تناول ما ينبغي في الشرع.

السفاهة: التسرّع إلى القول القبيح أو فعله.

الكذب: بيان[6] خلاف الواقع بالقصد.

الافتراء: اختراع قضية لا أصل لها.

الاختلاق: هكذا.

المين: بيانها لزيادة أو نقصان أو تحريف تغيّر[7] المعنى بها.

البذاء: الكلام القبيح.

---

[1] محبّة: محبّصة، ب. [2] لمن ينبغي: -، ب. [3] لجماعة: الجماعة، ب. [4] مشاركة الأصدقاء: مشاركة الأصداق، آ؛ مشاركه الأصدق، ب. [5] تصوّر: قصور، ب. [6] بيان: -، آ س. [7] تحريف تغيّر: تحريف تغيرت، آ؛ تحريف تغيّرت، ب؛ تخويف تغيرات، س.

الرفث: فواحش الكلام في[1] النكاح وأوصاف النساء.

السبّ: القدح في نسب شخص أو نفسه أو بدنه أو فعله.

العداوة: تحرّي اغتيال الغير ومضادّته فيما يؤدّي إلى مصالحه.

المكر: ما يقصد فاعله في باطنه خلاف ما يقتضيه[2] ظاهره.

الخديعة: ترادفه.

الظلم: وضع الشيء في غير موضعه.

الجور:[3] الخروج عن الوسط[4] بزيادة أو نقصان.

الكيد: إرادة متضمّنة لإستتار ما يراد عمن يراد[5] به.

الحسد: تمنّي خير يصل إلى غيره مع زواله عنه.

المنافسة:[6] السعي في أن يبلغ إلى مثل خير يصل إلى غيره.

فصل

العدالة: اسم يجمع سائر الفضائل.

الصداقة: محبّة صادقة لا يشوبها غرض بحيث يريد له ما[7] يريده لنفسه ويؤثّره على نفسه في الخيرات.

الألفة: اتّفاق الآراء في المعاونة على تدبير المعاش.

الوفاء: ملازمة[8] طريق المواساة ومحافظة عهود الخلطاء.

التودّد: طلب مودّة الأكفاء بما يوجب ذلك.

المكافأة: مقابلة الإحسان بمثله وزيادة.

المروءة: اسم للأفعال والأخلاق التي تقبلها النفوس[9] السليمة.

الإنسانية: الفضائل النفسية المختصّة بالإنسان.

الحرّية: أفعال وأخلاق محمودة لا تستعبدها[10] المطامع والأغراض الدنية.

---

[1]في: فحسب، ب. [2]يقتضيه: يقضيه، س. [3]الجور: الجوز، ب. [4]الوسط: الوسطة، ب. [5]يراد: يرادفه، آ؛ يزاد، س. [6]المنافسة: المنافة، س. [7]ما: + في، س. [8]ملازمة: الملازمة، س. [9]النفوس: النفس، آ. [10]لا تستعبدها: لا تستبعبدها، ب.

الفتوّة:[1] ما يختصّ به الفتى من الفضائل الإنسانية.

الظرف: اجتماع عامّة الفضائل النفسية والبدنية والخارجية.

حسن الشركة: رعاية العدل في المعاملات.

حسن القضاء: ترك الندم في المنّ في المجازاة.

صلة الرحم: مشاركة ذوي[2] القرابات في الخيرات.

الشفقّة: صرف الهمّة إلى إزالة المكروه عن الناس.

الإصلاح: التوسّط بين الناس في الخصومات بما يدفعها.

المشاورة: استنباط الرأي من الغير فيما يعرض من مشكلات الأمور الجزئية.

الغبطة: تمنّي خير يصل إلى غيره أن يكون له مثله.

النصيحة: إخلاص المحبّة لغيره في إظهار ما فيه صلاحه.

التوكّل: ترك السعي فيما لا يسعه قدرة البشر.

التسليم: الانقياد لأمر الله وترك الاعتراض فيما لا يلائم.[3]

الرضا: طيب النفس فيما يصيبه ويفوته مع عدم التغيّر.

البرّ:[4] السعة في علم الحقّ[5] وفعل الخير.[6]

الإخلاص: أن يقصد بما[7] يفعله وجه الله متعرّياً[8] عن الالتفات إلى غيره.

التقوى: جعل النفس في وقاية من[9] سخط الله تعالى.

الإحسان: تحرّي الحسنى في الإيمان والإسلام.

والله أعلم.

---

[1] الفتوّة: الفنون، ب. [2] ذوي: ذي، آ. [3] لا يلائم: يلائم، ب. [4] البرّ: السير، آ ب س. [5] الحقّ: الخلق، ب.
[6] الخير: الغير، آ س. [7] بما: ما، آ ب س. [8] متعرّياً: متقرباً، آ. [9] من: -، آ س.

الباب الحادي والعشرون: في التصوّف

التصوّف: هو العلم بالأحوال[1] الموروثة من تصحيح الأعمال ظاهراً وباطناً.

الوقت: حادث متوهّم علّق حصوله على حادث متحقّق.

المقام: ما يتحقّق العبد[2] بمنازلته من الأدب مما يتوصّل إليه بنوع تصرّف.[3]

الحال: معنىً يرد على القلب من غير تعمّل واكتساب.

القبض: معنىً يعرض لاستشعار مكروه[4] حاصل في الوقت.

البسط: معنىً يعرض لاستشعار[5] محبوب حاصل في الوقت.

الهيبة: معنىً يعرض لاستشعار عظمة المشاهد.

الأنس: معنىً يعرض لاستشعار لطف المشاهد.

وقيل: ارتفاع الحشمة مع وجود الهيبة.

وقيل: أن يستأنس بالأذكار فيغيب عن رؤية الأغيار.

وقيل: انبساط المحبّ إلى المحبوب.

الاتّصال: أن ينفصل بسرّه[6] عما سوى[7] الله.

وقيل: مكاشفات القلوب ومشاهدات الأسرار.

وقيل: وصول السرّ إلى مقام الذهول.

وقيل: أن لا يشهد[8] غير الله ولا يدخل بسرّه خاطر غير الله.

التواجد: استدعاء الوجد بضرب من الاختيار.

وقيل: ظهور ما يجد[9] في باطنه على ظاهره.

الوجد: جاذب القلب من فزع أو غمّ أو رؤية معنىً من أحوال الآخرة أو كشف حالة[10] بين الله تعالى وبين العبد.

وقيل: لهب ينشر في الأسرار تضطرب الجوارح طرباً أو حزناً عند ذلك الوارد.

---

[1] الأحوال: الأحول، ب. [2] العبد: + بما في [؟]، آ. [3] تصرّف: -، ب. [4] مكروه: محبوب، س. [5] لاستشعار: الاستشعار، ب. [6] بسرّه: لسرّه، آ. [7] سوى: غير، ب. [8] لا يشهد: يشهد، ب. [9] يجد: يحدث، ب. [10] حالة: حاله، آ ب س.

 | DOI:10.1163/9789004423367_029

وقيل: هو بشارة[1] الحقّ بالترقّي إلى مقام مشاهداته.

الوجود: خمود البشرية عند ظهور سلطان الحقيقة.

الغلبة: حال تبدو للعبد لا يمكن معها ملاحظة السبب ولا مراعاة الأدب.

الجمع: أن تكون الهموم كلّها همّاً واحداً فيصير ذلك حالاً له.

وقيل: جمع[2] الأسرار بأن ليس منه بدّ وقهرها فيه، إذ لا شبه له ولا ضدّ.

التفرقة: أن يفرّق بين العبد وبين هومه في حظوظه بطلب مرافق وملاذّه.[3]

وقيل: التقرّب إليه بالأعمال المتفرّقة[4] فإذا شاهدوه مقرّباً إليهم فهو الجمع.

الفناء: سقوط الأوصاف المذمومة.

وقيل: أن يفني عنه الحظوظ فلا يكون له في شيء حظّ ويسقط عنه التمييز فناءاً عن الأشياء كلّها شغلاً بمن فني فيه.[5]

البقاء: قيام[6] الأوصاف المحمودة.

الغيبة: أن يغيب[7] عن حظّ نفسه فلا يراها.

وقيل: اشتغال الحسّ بالوارد مع غيبة القلب عن علم أحوال الخلق.

الحضور: غيبوبته عن الخلق لاستيلاء[8] ذكر الحقّ على قلبه.

الشهود: أن يشهد بما يشهد مستصغراً له معدوم الصفة لما غلب عليه من مشاهدة الحقّ.[9]

السكر: الغيبة عن تمييز الأشياء بوارد استيلاء سلطان الحال.

الصحو: العود إلى ترتيب الأفعال.

وقيل: الرجوع إلى الإحساس بعد الغيبة.

وقيل: اختيار[10] المؤلم من موافقة[11] الحقّ ووجدان اللذّة فيه.[12]

الذوق: نتيجة الكشوفات بورود الواردات.

الشرب: يرادفه.

الريّ:[13] نتيجة الكشوفات بدوام المواصلات.

المحو: رفع أوصاف العادّة.

---

[1]بشارة: بشارات، ب. [2]جمع: جميع، س. [3]ملاذه: ملاذة، آ س. [4]المتفرّقة: التفرقة، آ ب. [5]فيه: به، ب س. [6]قيام: + لُو، ب. [7]أن يغيب: أن يغيبه، آ س. [8]لاستيلاء: الاستيلاء، آ. [9]الحقّ: الخلق، ب. [10]اختيار: احتيام، آ. [11]موافقة: موافق، آ س. [12]اللذّة فيه: اللذلة، آ س. [13]الري: الرأي، آ؛ الدىّ، ب.

الإثبات: إقامة أحكام العبادة.

الاستتار: حيلولة البشرية بين الله تعالى[1] وبين العبد.

التجلّي: رفع حجب[2] البشرية.

التجريد: أن يجرّد ظاهره عن الأعراض وباطنه عن الأعواض.

التفريد: أن لا يرى نفسه فيما تأتي به.

وقيل: أن يتفرّد عن الأشكال[3] ويتفرّد في الأحوال، ويتوحّد في الأفعال.

وقيل: التجريد أن لا يَملك والتفريد أن لا يُملَك.

المحاضرة: حضور القلب بتواتر البرهان واستيلاء السلطان.[4]

المكاشفة: حضور القلب بتعيين[5] البيان بلا تأميل دليل وتطلّب سبيل.

المشاهدة: وجود الحقّ من غير بقاء تهمة.

اللوائح: أمارات[6] طلوع شموس المعارف يكون زوالها بسرعة.

اللوامع: تلك الأمارات تبقى زماناً.

الطوالع: أمارات لها دائمة المكث موقوفة على خطر الأفول.[7]

البواده: ما يفجأ القلب من[8] الغيب على سبيل الوهلة فرحاً أو ترحاً.

الهجوم: ما يرد على القلب بقوّة الوقت[9] من غير تصنّع.

التلوين: الارتقاء[10] والتدرّج من حال إلى حال مختلفة.

التمكين: دوام استيلاء سلطان الحقيقة.

القرب: الاتّصاف في دوام الأوقات بمراضيه.[11]

وقيل: هو أن يتدلّل[12] على المحبوب.

البعد:[13] التدنّس بمخالفته.[14]

الشريعة: التزام العبودية.

الحقيقة: مشاهدة الربوبية.

---

[1] تعالى: -، ب. [2] حجب: محب، ب. [3] الأشكال: الشكال، ب. [4] السلطان: سلطان، ب. [5] بتعيين: بتعين، آ س. [6] أمارات: أمارة، س. [7] الأفول: الأقوال، ب س. [8] من: عن، آ س. [9] الوقت: القلب، س. [10] الارتقاء: لارتقاء، ب. [11] بمراضيه: بمرايضته، ب. [12] يتدلّل: يتدلّک، آ ب س. [13] البعد: العبد، ب. [14] بمخالفته: بمخالفه، ب.

علم اليقين: ما يحصل بالنظر والاستدلال.

عين اليقين: ما يحصل بطريق الكشوف.

حقّ اليقين: حقيقة ما يشير إليه علم اليقين وعين اليقين.

النفس: ترويح للقلوب بلطائف الغيوب.

الخاطر: خطاب يرد على الضمير بإلقاء ملك أو شيطان.

الوارد: ما يرد على القلب من الخواطر المحمودة بلا تعمّل.

النفس: ما كان معلولاً[1] من أوصاف العبد، مذموماً من أخلاقه وأفعاله.

الروح: عين لطيف مودع في القالب، أجرى الله العادّة نخلق الحياة ما دام هو في القالب.

السرّ: لطيفة مودعة في قالب الإنسان، يحيل فيها المشاهدة.

سرّ السرّ: ما لا اطّلاع عليه إلّا للحقّ[2] سبحانه وتعالى.[3]

فصل

المريد: من يريد الله مقبلاً بكلّيته إليه.

المراد: من يجذبه الحقّ جذب القدرة فيكاشفه بالأحوال.

الإرادة: ترك ما عليه العادّة.

وقيل: نهوض القلب في طلب الربّ.

التوبة: الرجوع من مذموم الشرع إلى محموده.

المجاهدة: فطم[4] النفس عن المألوفات وحملها على خلاف هواها في عموم الأوقات.

العزلة: الانفراد عن الخلق طلباً للخير.

الحزن:[5] حصر النفس عن النهوض في[6] الطرب.

المراقبة: علم العبد باطّلاع الربّ، واستدامته لهذا العلم.

وقيل: مراعاة السرّ لملاحظته الحقّ مع كلّ لحظة ولفظة.

---

[1] معلولاً: مغلولاً، آ ب س. [2] للحقّ: الحقّ، آ ب س. [3] وتعالى: -، ب. [4] فطم: وحم، آ س؛ وظم، ب.
[5] الحزن: الحون، آ. [6] في: + طلب، آ.

وقيل: خلوص السرّ والعلانية لله.

التقوى: التحرّز بطاعة الله عن عقوبته.

وقيل: مشاهدة الأحوال على قدر الانفراد.

وقيل: ترك ما دون الله.

وقيل: مجانبة الهوى ومصاينة النفس.

الورع: ترك الشبهات.

الزهد: خلوّ[1] القلب واليد عن الإملاك والتتبّع.

الفقر: أن لا يطلب المعدوم حتّى يفقد[2] الموجود.

وقيل: أن لا يكون له فإذا كان لا يكون له.

وقيل: عدم كلّ موجود وترك كلّ مفقود.

الخوف: أن يخاف من نفسه أكثر مما يخاف من عدوّه.

وقيل: أن لا يخاف غير[3] الله.

الرجاء: ارتياح القلب برؤية[4] كرم المرجوّ.

وقيل: رؤية الجلال بعين الجمال.

وقيل: قرب القلب من ملاطفة الربّ.

التواضع: الاستسلام للحقّ وترك الاعتراض على الحكم.

وقيل: تذلّل القلوب لعلّام الغيوب.

وقيل: قبول الحقّ من الحقّ للحقّ.

وقيل: الافتخار[5] بالقلّة والاعتناق للذلّة وتحمّل أثقال أهل الملّة.

الخشوع: الانقياد للحقّ.

وقيل: خفض الجناح وكسر الجانب.

الحياء: وجود الهيبة[6] في القلب مع وحشة[7] ما سبق.

وقيل: ذوبان الحشا لاطّلاع المولى.

وقيل: انقباض القلب لتعظيم الربّ.

---

[1] خلوّ: خلق، آ. [2] يفقد: يفقل، ب. [3] غير: -، س. [4] برؤية: تروية، ب. [5] الافتخار: افتخار، ب.

[6] الهيبة: الهيئة، آ. [7] وحشة: خشة، آ.

القناعة: ترك التشوّق[1] إلى المفقود والاستغناء بالموجود.

التوكّل: ترك تدبير النفس والانخلاع عن الحول والقوّة.

وقيل: أن يكون لله كما لم يكن فيكون[2] الله كما لم يزل.

وقيل: الاستسلام لجريان القضاء في الأحكام.

وقيل: الاسترسال بين يدي الله عزّ وجلّ.[3]

وقيل: ترك الأدواء[4] إلى الله.

وقيل: أن لا يطالب الله بالأعواض.

الصبر: الوقوف على البلاء بحسن الأدب.

وقيل: التباعد عن المخالفات، والسكون عند تجرّع غصص البليّات، وإظهار الغنى عند حلول الفقر بالساحات.

وقيل: الغنى في البلوى بلا ظهور شكوى.

وقيل: المقام مع البلاء بحسن الصحبة.

وقيل: أن لا يفرّق بين حالي النعمة والمحنة مع سكون الخاطر فيهما.

الشكر: الاعتراف بنعمة المنعم على وجه الخضوع.

وقيل: مشاهدة المنّة وحفظ الحرمة.

وقيل: أن لا ترى نفسك أهلاً للنعمة.

وقيل: الغيبة عن الشكر برؤية المنعم.

الشاكر: من يشكر على الموجود.

الشكور: من يشكر على المفقود.

وقيل: الشاكر من[5] يشكر على الرفد، والشكور من يشكر على الردّ.

وقيل: الشاكر من يشكر على العطاء، والشكور من يشكر على البلاء. ويقال: الشاكر من يشكر عند البذل، والشكور من يشكر على المطل.

الرضا: رفع الاختيار.

وقيل: سكون القلب تحت جريان الحكم.

---

[1] التشوّق: التشوّف، ب. [2] فيكون: فيكو، س. [3] عزّ وجلّ: -، س. [4] الأدواء: الأبواء، آ؛ الايواء، ب س. [5] من: -، آ س.

وقيل: سرور القلب بمرّ القضاء.[1]

وقيل: استقبال الأحكام بالفرح.

وقيل: نظر القلب إلى قديم اختيار الله للعبد.[2]

اليقين: ارتفاع الشكّ.

وقيل: ما زالت عنه المعارضة على دوام الوقت.

وقيل: اتّصال البين[3] [وانفصال ما] بين البين.

المحبّة: هي تعظيم محلّ الأسرار.

وقيل: ميل القلب إلى الله وإلى ما لله من غير تكلّف.

وقيل: الطاعة فيما أمر والانتهاء عما زجر والرضا بما حكم وقدر.

العبودية: التبرّي من[4] الحول والقوّة والإقرار بما يوليك[5] من الطول والمنّة.

وقيل: القيام بحقّ[6] الطاعات بشرط التوقير والنظر إلى ما منك بعين التقصير وشهود[7] ما يحصل من مناقبك من التقدير.

وقيل: ترك الاختيار فيما يبدو من الأقدار.[8]

وقيل: إسقاط رؤية التعبّد في مشاهدة المعبود.

وقيل: العبادة لمن له علم اليقين والعبودية لمن له عين اليقين والعبودة[9] لمن له حقّ اليقين.

وقيل: العبادة لأصحاب المجاهدات[10] والعبودية لأرباب المكابرات والعبودة[11] صفة أهل المكاشفات.

الإخلاص: إفراد الحقّ في الطاعة بالقصد.

وقيل: ارتفاع الرؤية عن الفعل.

وقيل: تصفية العمل عن ملاحظة المخلوقين.

الاستقامة: أن يشهد الوقت قيامه.

وقيل: الخروج عن المعهودات ومخالفة[12] الرسوم والعادّات والقيام بين يدي الله على حقيقة الصدق.

الحرّية: أن لا يكون تحت رقّ المخلوقات، ولا يجري عليه سلطان المكونات.

---

[1] القضاء: الرضاء، آ س. [2] للعبد: + الا، س. [3] البين: الهين ما، ب. [4] من: عن، آ ب س. [5] يوليك: اوليک، آ س. [6] بحقّ: الحقّ، آ؛ لحقّ، س. [7] شهود: الشهود، ب. [8] الأنوار: الاقرار، آ ب س. [9] العبودة: العبودية، س. [10] المجاهدات: المشاهدات، آ. [11] العبودة: العبودية، آ س. [12] مخالفة: المخالفة، آ.

الفتوّة: أن يكون أبداً في أمر غيره لله[1] تعالى.

وقيل: هي الصفح عن عثرات الإخوان.

وقيل: كفّ الأذى وبذل الندى.

وقيل: أن لا يرى لنفسه فضلاً على غيره.

وقيل: أن تكون خصماً لربّك على نفسك.

وقيل: أن يَنصف ولا يُنصَف.[2]

وقيل: أن لا تنافر فقيراً ولا تعارض غنيّاً.

الصدق: موافقة السرّ النطق ومنع الحرام من الشدق.

وقيل: استواء السرّ والعلانية.

وقيل: صحّة التوحيد مع القصد.

التوحيد: إفرادك متوحّداً، وهو أن لا يشهد[3] الحقّ إيّاك لك.

وقيل: إفراد الموحّد بتحقيق وحدانيته بكمال أحديته.

وقيل: معنىً يضمحلّ فيه الرسوم، ويندرج فيه العلوم، ويكون الله كما لم يزل.[4]

وقيل: إسقاط الوسائط عند غلبة الأحوال والرجوع إليها عند الأحكام.

وقيل: هو أن يرجع العبد إلى أوّله فيكون كما كان قبل أن يكون.

الموحّد: أن يستوحش من سرّه وحشة لظهور الحقّ عليه.

وقيل: هو من حال الله بينه وبين الدارين جميعاً.

وقيل: أن لا يجري عليه ذكر أخطار ما[5] لا حقيقة له عند الحقّ، فالشواهد عن سرّه مصروفة، والأعواض عن قلبه مطرودة.

العارف: من بدا له الشاهد، وفني الشواهد، وذهب الحواسّ[6]، واضمحلّ الإخلاص.

وقيل: أن يكون كما[7] كان حيث كان قبل أن يكون.[8]

وقيل: من بذل مجهوده[9] فيما لله ويحقّق معرفته بما منّ الله[10] وصحّ رجوعه من الأشياء إلى الله.

الوليّ: من يتولّى الحقّ رعايته.

---

[1] لله: الله، آ. [2] أن ينصف ولا ينصف: أن يتصف ولا يتصف، آ ب س. [3] يشهد: تشهد، س. [4] لم يزل: أنزل، آ. [5] أخطار ما: اخطارها، آ ب س. [6] الحواس: الخواص، آ ب س. [7] كما: لما، آ س. [8] قبل أن يكون: -، آ س. [9] مجهوده: محمود، ب. [10] ويحقّق معرفته بما منّ الله: -، آ س.

وقيل: من توالت أفعاله على الموافقة.

وقيل: هو الفاني عن حاله، الباقي في مشاهدة ربّه.[1]

الفراسة: خاطر يهجم على القلب فينفي[2] ما يضادّه.

وقيل: سواطع أنوار لمعت في القلوب.

وقيل: مكاشفة اليقين ومعاينة الغيب.

والله أعلم.[3]

[1] من الأشياء إلى الله [...] في مشاهدة ربّه: كتبت في هامش ب. [2] فينفي: فيبقى، آ. [3] أعلم: + حقيقة كلّ شيء، آ؛ + حقيقة كلّ شيء تمّت سنة 1181 ماه جمادى الآخر، س.

فهرس المصطلحات